CONTEMPORARY'S

The GED Math Problem Solver

TEACHER'S GUIDE

MYRNA MANLY

Project Editor
Kathy Osmus

CB
CONTEMPORARY
BOOKS
CHICAGO

Library of Congress Cataloging-in-Publication Data

Manly, Myrna.
The GED math problem solver : teacher's guide / Myrna Manly.
p. cm.
ISBN 0-8092-4049-1 (paper)
1. Mathematics—Problems, exercises, etc. 2. General educational development tests. I. Title.
QA43.M26 1992
510'.76—dc20

91-42166
CIP

Published by Contemporary Books, Inc.
Two Prudential Plaza, Chicago, Illinois 60601-6790
Manufactured in the United States of America
International Standard Book Number: 0-8092-4049-1

Published simultaneously in Canada by
Fitzhenry & Whiteside
195 Allstate Parkway
Markham, Ontario L3R 4T8
Canada

Editorial Director
Caren Van Slyke

Editorial
Karen Schenkenfelder
Erica Pochis
Claudia Allen
Janet Fenn
Esther Johns
Eunice Hoshizaki

Editorial Production Manager
Norma Fioretti

Cover Design
Georgene Sainati

Art & Production
Carolyn Hopp

Interior Graphics
J•B Typesetting
St. Charles, Illinois

Typography
Ellen Kollmon

Front cover computer-generated graphic by Klug Präzis

Contents

Section Four: **Ratio, Proportion, Percent, and Data Analysis**

Handouts: **Handouts 1 Through 6**

Introduction

The GED Mathematics Test

The GED Math Test is constructed to test problem-solving skills using real-life contexts. Marketplace, workplace, and domestic situations provide the settings for problems that require a variety of mathematical skills. To create these real-life settings, the test is made up of "word problems," which the examinee must interpret as mathematical problems. In some of the problems, examinees are asked only to set up the solution, not to find the final answer.

The GED Math Problem Solver and this teacher's guide present a curriculum based on the specifications of the GED Test. Thus, they provide an efficient, up-to-date course of study that prepares students for the kinds of mathematical problems that will confront them in everyday life and, therefore, on the test.

Teacher Preparation

Before you attempt to prepare students to pass the GED Math Test, you should have a general idea of what the test requires. The Official Practice Tests (authorized by the GED Testing Service, One Dupont Circle, N.W., Washington, D.C. 20036) are the best indicators of what the actual test is like, and they should be available in your area. If possible, try each of the practice forms (currently, there are four forms—AA, BB, CC, and DD).

As you take the tests, keep the following points in mind:

- Note how the test deemphasizes computation skills by using the set-up format and "nice" numbers.
- Try estimating the answers before you do any paper-and-pencil computation. Note how often this strategy is adequate for choosing a response. Even a beginning estimator could choose the correct answer for about 75 percent of these problems without using paper-and-pencil algorithms.
- Notice that the decimal and fraction skills needed are only those required by everyday experiences.
- Note that the algebra problems use only the most fundamental concepts of what could be called numerical algebra. Many of the items that are classified as algebra can be solved by arithmetic methods.
- Notice the frequency of problems that can be solved using ratio and proportion. Often as many as 40 percent of the test questions involve an understanding of these concepts.

Implications for Instruction

Once you have taken the practice tests yourself, you will have a better feel for what is required on the actual GED Test. Students can meet the test requirements if you follow these guidelines:

- Concentrate on problem analysis and interpretation.
- Attempt to narrow the gap between school math and real-life math.
- Integrate the use of calculators into the curriculum.
- Teach techniques involved in estimation and mental computation.

- Use algebra to generalize the basics of arithmetic as they are learned.
- Use geometry topics to illustrate applications of arithmetic concepts.
- Stress understanding of concepts rather than rote manipulation procedures. Also deemphasize finding the answer—focus on process.

Important Skills for Students

Throughout the book, students will have to be able to read analytically in order to understand the problems. The book teaches estimation strategies that students can apply to their daily life as well as the GED Math Test. When applied to estimating, some math skills become more important than others. The following skills recur throughout the book:

- Single-digit facts
- Place value and × and ÷ by powers of 10
- Rounding to adapt the problem
- Fraction/decimal equivalents
- Confidence in fundamental math principles

As students progress, stress these skills as the necessary basics for succeeding with the unique approach of the book and, therefore, the GED Math Test.

Calculators in the Classroom

Calculating machines, especially cash registers and personal calculators, have become an integral part of our lives. Students should have the opportunity to learn how to use these machines intelligently.

Many of our adult students suffer from math anxiety. These fears will be calmed if they are able to use calculators in class. Since many students have not mastered the traditional computational algorithms, using calculators will provide them with the freedom to reason with numbers and solve everyday problems. Students can become functionally math-literate.

In adult education classes, students have widely different skill levels. The calculator can serve as the "great equalizer." Then you can dedicate the available class time to teaching problem-solving techniques, rather than concentrating on rote computation methods. Research has shown that students who have used calculators in the classroom perform better on tests—even when calculators are not allowed while taking the tests. The students who used calculators in the classroom improved their problem-solving skills dramatically during these research periods.

Throughout the book, students will be instructed to use their calculator on any problem for which they think it appropriate. Students should soon discover that calculators are necessary only when tedious computation is required or a precise answer is needed. For most problems, students will find solutions through estimation, mental math, or simple paper-and-pencil methods. In this way, students come to realize that calculators would not be very helpful to them on the GED Test, where complicated computations are minimal. Point this out to students who worry that they will become dependent on the calculator and thus be at a disadvantage. (At the time of this printing, GED examinees may not use a calculator.)

Structure of the Student Book

Lesson Design

There are 28 lessons divided into four sections by subject area. Each lesson is designed to fill a 90-minute class period. However, your class can progress at a rate that is comfortable for your students.

Mental Math Exercises

The exercises given at the beginning of most of the lessons can serve as warm-ups for your students. Sometimes they will solidify everyone's thoughts by reviewing concepts from the previous lesson. At other times, they will introduce the ideas being presented in the lesson.

Lively discussions will certainly result if you encourage students to share their strategies as well as their answers to the exercises. You will need to devise a system that gives all students adequate time to solve the problem before the quicker students blurt out the answer. If you ask them to write only the answer on a piece of paper while you walk around the room, each individual will be encouraged to participate. You could introduce the use of hand signals so that students can indicate when they are ready with the answer, and then they can show their answers in unison. Feel free to find a method that best suits your class's personality and needs.

Principles, Examples, and Problems

The learning activities that make up the bulk of each lesson have been designed to serve as the skeleton for group work. Student interaction is vital to the success of this approach. Students need to verbalize their understandings. In this way, they are able to learn from each other.

Find a way for students to do the problems interspersed throughout the lesson in a group situation so that they can also experience doing math together. The problems are structured so that students, with your guidance, can discover the concepts basic to the lesson. Guided discovery is the best description of what will be going on in your classroom. Listen carefully to student comments: they are vital to your making the instruction fit the needs of your particular class.

Check Your Understanding

The problems at the end of each lesson provide students with the opportunity to independently solve new but related problems. It would be best to give students enough time at the end of the session to solve these problems. At times the Check Your Understanding section could be considered homework, but this approach is less likely to be successful.

Answer Key

The answers in the answer key are meant to be part of the learning process. Students aren't required to use the same solution methods given in the answer key, but they can learn by comparing their methods with the ones shown.

Structure of the Teacher's Guide

This guide gives you additional materials to flesh out your instruction as well as suggestions for supplemental exercises.

Objectives

The objectives give an overall view of what students will be accomplishing in the lesson.

Background

Background information for each lesson will help you and your students understand what direction the lesson takes and reasons for that particular approach.

Lesson Recommendations and Extension Activities

The recommendations for each lesson give a starting point for discussion and ideas for different solution methods. Also included in this section of the teacher's guide are extension activities designed to help students discover and/or reinforce various principles of mathematics.

• Group Learning

As successful adult students recount their learning experiences, they often mention *other people* as the deciding factor in their success. As the teacher in an adult education center, you are most likely going to be one of the people who make a difference in a student's life. But if your job at the center is merely to pass out materials and record student progress, this likelihood diminishes. Students on the edge need conversations about topics that have been foreign to them. They need to belong to a group that stimulates their curiosity and helps them gain the satisfaction that comes from learning.

This is only one of the reasons why group instruction is often preferred over an individualized approach for adult remedial math instruction. In a group, students identify with each other and gain comfort in knowing they are not alone. They encourage each other as they make small steps toward successfully completing the course. The classroom interaction also gives them an opportunity to contribute their ideas for solution strategies they have learned from experience, thereby enhancing their self-esteem.

In summary, the socialization that group learning provides is important. It will keep your students coming back each session and provide you with the opportunity to inject some fun into the learning process. For me as a teacher, it also brings back the feelings of personal satisfaction that caused me to enter the profession in the first place.

• Recommendations for the First Class

GED Test Preview

Think of the GED Test Preview as an introductory packet to the course. Some students will read it before they meet you, and others will be introduced to you and the course by your discussion of it. This form of nonthreatening exposure to the kinds of problems that students can expect to find on the test is preferable to a required pre-test. As mentioned earlier, the ability to read analytically and a working knowledge of the single-digit math facts are the most important prerequisites for this material.

Note: If your center requires a pre-test for some reason other than instructional purposes, try to ensure that the circumstances surrounding the pre-test remain as nonthreatening as possible. Remember that your class includes many fragile learners who may be put off by a test that asks about things they do not know.

In addition to making students more aware of what is ahead of them when they take the GED Test, this preview also validates the book's approach to learning mathematics. It will be a radical departure from what most students expect from a remedial mathematics course. They should know from the outset that this innovative approach will serve them well in terms of learning what they need to know to pass the test. It provides an efficient course of study covering the topics most likely to occur on the test.

Introductions

Spend the entire first session on introductions—to each other and to the course. Encourage students to talk about why they have come to the class. Discuss the practical uses of math. To stimulate conversation, you can say, "Tell us about a time in your life when you needed mathematics in an everyday situation," or, "Describe a situation where you intend to use the math you learn in this course." You can prompt students by asking about numbers in sports, cooking, home or car maintenance, or a hobby. Listen carefully and note these situations. Bring them up again later, using them as motivation when you are studying the appropriate skills. Nothing is as meaningful as this dilemma-driven learning.

Next, work through the GED Test Preview in the student book. The problems chosen for this preview not only illustrate various aspects of the test, but also represent the goals your students should aim for in their studies. No one is expected to know how to do these problems at this time. Read and discuss each problem so all of the students understand what is being presented. Then discuss the various answer choices as they are listed in the book. These problems are not meant as learning activities, but as *awareness builders*.

Discussion of Calculator Use

Explain to students that using calculators will play an important role in learning to use mathematics. Encourage them to buy a simple, solar-powered calculator so they can become comfortable using it. The calculators need only be capable of the four operations: $+$, $-$, $\times$, and $\div$. However, the following keys also will be mentioned: [M+], [M−], [MR], [%], and [$\sqrt{\ }$].

Attendance

Encourage attendance as much as possible. The lack of this important life skill may be one of the reasons that students need this "second chance." Stress that the class will be depending on each individual to make the discussions real. At first, some may be willing only to listen and observe. Make these students feel important each session by a cheerful greeting, an inquiry about their progress or attitude, or an offer to give special assistance. Let them know that everyone has something important to contribute.

If it is possible at your learning center, offer times between class sessions for individual tutoring. This will give the slower learners a better chance of keeping up with the others.

GED Readiness

You may have some students who, after seeing what the test is like, feel they can pass the test without further instruction. Have these students try an Official Practice Test, the only true predictor of success on the GED Test. If a student scores high enough to merit an immediate try at the test, you will have done him or her a great service. If the score shows a need for more instruction, you will have gained a more motivated student.

LESSON 1 • "Seeing" Addition and Subtraction

Objectives

In this lesson, students will

- develop intuition for deciding from the context of real-life problems when to add or subtract; learn to respond to clue relationships rather than clue words;
- use variables (letters in place of numbers);
- use mathematical notation to describe a problem;
- use mental computation techniques based on the knowledge of basic facts and place value;
- by the end of the addition and subtraction section, be able to state sums and differences of single-digit numbers (using flash cards or computer drill) within 3 seconds.

Background

Addition and Subtraction Facts

As students progress through this lesson, you will detect areas of weakness in some students' basic addition and subtraction skills. This is normal and simply reflects a need to practice and use those skills.

EXTENSION ACTIVITY: **BASIC FACTS**

Handout 1
BASIC FACTS
TG Page 101

In a one-on-one tutoring session, you can assist students who have difficulties with basic addition and subtraction by helping them to construct an addition table like the one on page 101 of this guide.

1. To construct the table, have the student label the row and column headings from 0 to 9. Then have the student fill in the answers that he or she knows. After doing that, make sure the student is aware of the following points:
   - Adding 0 does not change the number. (Point out the first row and column.)
   - Adding 1 is just like counting; find the next number. (Refer to the second row and column.)
   - Adding 2 is just adding 1 twice.
2. It is not necessary to fill in the answers in any particular order. In fact, by asking the student to fill in only those answers that he or she knows, you can diagnose the student's difficulties.

3. As you continue to complete the table together, help the student recognize the patterns formed because of commutativity ($4 + 5$ is the same as $5 + 4$). Talk about odds and evens (odd + odd = even, even + even = even, odd + even = odd). Note the sums that are equal to 10 and the ones larger than 10. And note the "doubles" of each number on the diagonal of the table.

4. If a student can isolate where he or she is having difficulties, show some compensatory ways to figure out the sum in case he or she forgets the answer. For example, if the student has trouble with $7 + 8 = 15$, you can refer to more familiar facts: $7 + 7$ and $8 + 8$. The answer to $7 + 8$ falls in between.

5. When the table is completed correctly, allow the student to use it whenever he or she needs it.

• Lesson Recommendations

Pages **2-3**

1. Picture the Situation

At the beginning of the lesson, the students use pictures to help them visualize the action involved when they add or subtract. To introduce addition, emphasize the actions involved with *combine* and *join*, and use descriptive phrases like *bring together* as you discuss the problems.

Subtraction is introduced here through the actions of *separating* and *comparing*. The clue relationships will be the basis for deciding which operation to use in the problem. Determining this relationship will be pivotal to the most critical step of problem solving: understanding the problem.

Pages **4-5**

2. Write the Problem

The step of writing the problem is stressed in this book for many reasons:

- It deemphasizes the focus on finding the answer.
- It prepares students for the use of variables.
- It enables students to choose the correct response in set-up problems on the GED Test.
- It provides a stopping place to decide how to proceed to the answer.

Many students resist doing this step. They want the answer, and nothing else seems important to them. You must persist in helping them overcome this tendency. For one thing, it is self-defeating. If students do not understand or set up a problem correctly, they will not be able to get the correct answer.

You can also use this opportunity to ease students away from the vertical to the horizontal format of writing problems. When written horizontally, the problem will be in the format in which mathematics is most often communicated, not only in writing, but also when being entered into a calculator or computer.

Variables are used in everyday experiences. The algebra in the student book is straightforward and reflects the arithmetic it generalizes. If the students are introduced to algebraic concepts gradually, they will not have any trouble accepting them. Make a point of showing how the examples with variables use the same words as the examples with numbers **(Problem 6)**.

Page **6**

3. Find the Answer

For many years, mathematics instruction has focused on one step of the problem-solving process: finding the answer. Therefore, it is no surprise that the students in your class think that is the most important task in math. They want to jump to the answer even before reading the whole problem.

Encourage students to consistently focus on the prior two steps of *understanding* and *writing* the problem. Finding the answer is only one step of the process, and since students may also use calculators, it is not the step the student book emphasizes.

Sometimes the method chosen to find an answer depends on the situation and at other times on the person solving the problem. For example, the problem itself may not require a precise numerical answer (as in the case of finding the amount of paint to buy for a room). On the other hand, a shopper in a department store would *estimate* the price of an item that is marked 25% off but would not allow the cashier to estimate the price at the cash register.

Estimation and mental computation should be considered as the first options when trying to find the solution to a problem. They are the quickest and most efficient methods available. They are especially appropriate for a class of adults. Students' self-esteem will grow each week as they see what powerful tools they already have.

On page 45 of the student book, a flowchart shows the process that this book advocates for solving the problems on the GED Math Test.

Pages **6-7**

Adding and Subtracting Mentally

The first mental math problems in this lesson **(Problem 7)** ask the student to concentrate on the first digit of each number and to use basic addition and subtraction facts ($2 + 3 = 5$, $20 + 30 = 50$, etc.). Most students will not require any additional explanation of these. However, the set requiring them to focus on the end digit **(Problem 9)** should be explained by using one or both exercises in the following activity.

EXTENSION ACTIVITY: **HUNDREDS CHART**

Handout 2
Hundreds Chart
TG Page 102

Make a transparency of the hundreds chart (TG page 102) to project on a screen, or give each student a copy to follow. The purpose of this activity is to show through patterns what is involved when you "carry" in addition and "borrow" in subtraction. (Carrying and borrowing are sometimes called **regrouping**.) The process remains the same, even though the numbers to which you are adding differ. The first column of problems **(Problem 9a)** starts with $6 + 8 = 14$.

Problem 9a

1	2	3	4	5	6	7	8	9	10
11	12	13	14	15	16	17	18	19	20
21	22	23	24	25	26	27	28	29	30
31	32	33	34	35	36	37	38	39	40
41	42	43	44	45	46	47	48	49	50
51	52	53	54	55	56	57	58	59	60
61	62	63	64	65	66	67	68	69	70
71	72	73	74	75	76	77	78	79	80
81	82	83	84	85	86	87	88	89	90
91	92	93	94	95	96	97	98	99	100

1. Circle the 6.
2. Count 8 places, landing on the 14.
3. Put a box around 14.

Repeat the process with 36 + 8 and again with 56 + 8. Ask students to notice the pattern: the numbers being boxed are in the next row from the circled numbers (a jump to the next tens digit), and they all end in 4 (which is 2 less than 6). The 2 from the "2 less than 6" combines with the 8 in the problem to make 10. By the time students get to 126 + 8, they won't need the chart.

Students may already know the pattern that occurs when 9 is added to a number. If so, ask a student to illustrate it on the chart for **Problem 9c**.

Problem 9b

1	2	3	4	5	6	7	8	9	10
11	12	13	14	15	16	17	18	19	20
21	22	23	24	25	26	27	28	29	30
31	32	33	34	35	36	37	38	39	40
41	42	43	44	45	46	47	48	49	50
51	52	53	54	55	56	57	58	59	60
61	62	63	64	65	66	67	68	69	70
71	72	73	74	75	76	77	78	79	80
81	82	83	84	85	86	87	88	89	90
91	92	93	94	95	96	97	98	99	100

Similarly, borrowing is demonstrated by this visual aid for 11 − 7 = 4.

1. Circle 11.
2. Count 7 places back to subtract.
3. Put a box around 4.

Repeat for 31 − 7. Ask the students to describe the pattern being shown.

You may encounter students who have developed a mental block about the carrying and borrowing processes. Alternative ways to compute are shown in the appendix of the student book (page 247). These alternative ways to add and subtract are targeted for the students who are having problems. However, your other students will also enjoy learning additional ways to find the answers.

Page **8**

Rounding and Estimating

The estimating that is covered in this lesson is called, appropriately, **front-end estimation**. Intentionally, no rules were introduced for rounding. Students are often able to see what each number is "closest to" without being bogged down by rules.

When students are working through the example, ask questions such as these: Should 367 be placed closer to 300 or 400 on the number line? Which number is exactly halfway between them? Encourage the kind of commonsense thinking that will tell students how to round in these cases.

Problem 10 includes numbers that involve the transition between 900 and 1000. It would be helpful to use the expressions "ten hundred" and "eleven hundred" for these positions on the number line of hundreds. When discussing how the place-value system works, show how these expressions equal one thousand and one thousand, one hundred, respectively.

The same concept arises again with the number line students use in **Problem 11**.

LESSON 2 • Grouping to Add More than Two Numbers

Objectives

This lesson will help students to

- use the associative law and parentheses to group addends;
- evaluate algebraic expressions;
- estimate sums using compatible pairs and grouping;
- use calculators to find precise sums;
- recognize the relative size of decimal numerals.

Background

Flexibility in Adding Numbers

Flexibility is important to problem solving. Students must be able to use a variety of methods to find the answer to a problem, especially when estimating. To be flexible, however, your students must develop confidence with respect to the fundamental principles of math. For example, if students are certain that they can change the order and grouping of numbers when they add, they will see that there are many ways to approach problems.

For this reason, the **commutative** and **associative** properties of addition are explained and then used to find precise and approximate answers. In addition, compatible pairs of numbers give students easy numbers to work with. For example, numbers that add to 10 are compatible, as are numbers that add to 100. (This is especially important when dealing with money.) All of these ideas are combined in this lesson, where grouping is taught as a technique for adding mentally.

Lesson Recommendations

Page **10**

Mental Math Exercises

All of the lessons (excluding checkpoints) that follow will be preceded by a set of mental math exercises that are a warm-up to the lesson topic or a quick review of an earlier lesson. If your students respond well to these, you may wish to construct your own groups of problems to supplement those given. For example, you could extend this set to include pairs of numbers that add to 100. Try to group the problems so that they have a common theme or work up to a concept.

Pages **10–11**

Perimeters

Perimeters offer a concrete application for adding more than two numbers. With perimeters, you can teach students to use the associative law by grouping the addends into compatible pairs. (Students are not required to know the names of the laws, only when to apply them.)

Try to convince your students to take time to analyze a problem first, instead of immediately beginning to add from left to right. They should think about how to regroup the numbers to make adding easier. The importance of thinking about the problem before jumping into finding the answer will be a recurring theme in this book.

Page **11**

Grouping with Variables

This section reintroduces the concept of variables from **Lesson 1**. The two perimeter examples show how to group the numbers, add them, and indicate the remaining addition to be done in the expression.

Page **12**

Evaluating Expressions

The valuable algebraic skill of evaluating expressions is introduced here as a follow-up to perimeters. The word *substitute* is important. To promote the idea of a variable, you may say, "Use 16 instead of the letter x," or, "Substitute 16 for the letter x."

EXTENSION ACTIVITY: **UNDERSTANDING x**

1. Put different numbers of strips of colored paper into envelopes. Seal each envelope, and write an x on it.
2. Give each student (or group) one envelope and four additional strips of paper. Each of them now has $x + 4$ strips of paper. Write this expression on the board.
3. Ask each student or group to open its envelope and determine the value of x. Remind the students that each student or group has a different value for x. Then ask, "If you add four to the value you have for x, how many strips of paper do you have in all?"
4. Now write out each equation. For example, if someone has 5 strips in the envelope, write $x + 4 = 9$. Then have that student or group plug in the value for x as reinforcement.

Pages **12-13**

How Much Is the Total Cost? Estimating and Calculating

This section focuses on learning math for the way we use it. Many situations in our lives do not require an exact total. If the precise answer is needed, either a cash register or a calculator can provide it. What is important for most of us is to be able to find an approximate total to provide a quick check against carelessness.

The only real rule of estimating is that you must simplify the problem enough to be able to do it in your head. The students are sharpening their mental skills so that they can feel comfortable with estimating and mental calculating.

In **front-end estimation**, focus on the digit that tells the most about the size of a number—the left-most digit. Sometimes an estimate needs to be refined by first **rounding** or **grouping** the cents into dollars. Rounding is more appropriate when the values are near the extremes ($1.99 or $4.07) **(Problem 8)**, and grouping works better with intermediate values (when the cents add up to approximately a dollar, as in **Problem 9**).

We teach these different methods of estimation so that students can choose the one that will give a more accurate answer. However, no one estimate is the only right estimate. Many students will have developed strategies of their own to apply to these situations, and some of their strategies may be more practical than those presented here. Be sure to maintain a classroom atmosphere where all reasonable estimates are acceptable, and encourage explanations of unique methods.

Page **14**

Money and Decimals

Students often have trouble understanding the concept of a decimal, but they know about money. Use this knowledge to help students make the connection between money and pure decimals. An important point to stress here is how the cent values ending in zero are equal to a pure decimal without the zero ($.40 = .4, $.30 = .3). (When a number is expressed as a pure decimal, the trailing zero is not necessary.)

EXTENSION ACTIVITY: **DECIMALS IN MONEY**

Show various sums of money to students, and ask them to write the values in decimal notation. The point of this activity is to show the students that from their experience with money they already know

- the significance of the decimal point;
- a numbering system with tenths and hundredths. (One cent is one-hundredth of a dollar.)

Page **15**

Comparing Decimal Numbers

To compare decimal values, students need to understand which zeros in a decimal number are important in determining the value and which are not. For example, in .40, the zero does not change the value of the decimal, since .40 = .4. But in .04, the zero is critical to determining the value of that number, since .04 is not equal to .4.

Be certain that students know where and when they can add zeros to a decimal without changing its value.

EXTENSION ACTIVITY: **UNDERSTANDING DECIMAL PLACES**

Use the charts found on page 107 of this book to show students that adding zeros does not change the value of a decimal and can be used to give decimal numbers an equal number of digits.

1. Make two separate transparencies of the tenths grid and the hundredths grid or cut one transparency of both grids in half.

2. On an overhead projector, superimpose one grid over the other to show that the entire square of each grid equals one.

3. Illustrate the comparisons in **Problem 11**. For example, in **11a**, shade the correct amounts on the appropriate grids, superimpose, and compare. Then show that shading sixty on the hundredths grid is the same as shading six on the tenths grid.

tenths grid

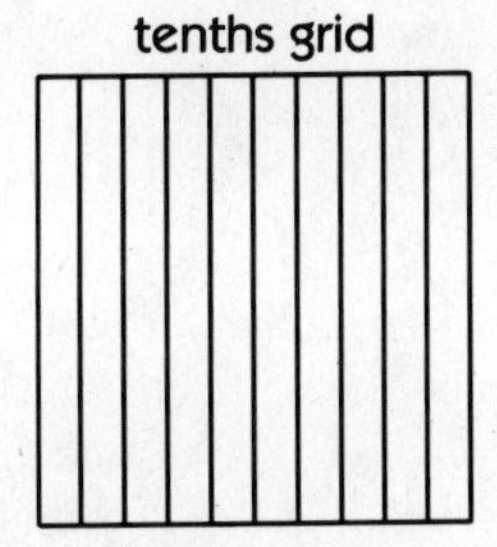

hundredths grid

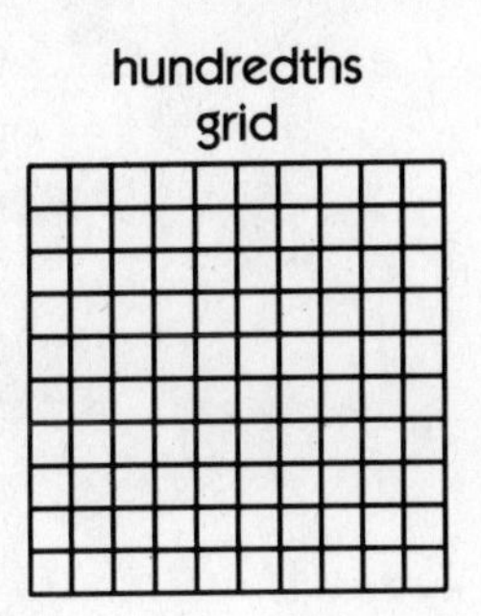

EXTENSION ACTIVITY: **UNDERSTANDING ZEROS IN DECIMALS**

To show whether the zeros in a decimal number are critical to its value, have students do the following activity with their calculators. Please note that this activity will not work with all calculators.

a) 6.40 64.00 6.04

b) 1.50 1.05

c) 46.20 46.02

d) 12.00 1.20 1.02

1. Have students enter each number separately in their calculator, followed by an equals sign. (Remind students to clear their calculator before each entry.)

2. Have students notice what number is displayed on their calculator each time. Is the 0 retained? If yes, the 0 is critical. If not, it is not critical.

Page **16**

Reading Word Problems

At this juncture, some students will benefit from this exercise, which will help them to read word problems more analytically. By asking the simplest question, "Who went farther?" **(Problem 15)**, students will learn to focus carefully on the question that is being asked.

EXTENSION ACTIVITY: **CALCULATING AND ESTIMATING WITH A CHECKING ACCOUNT**

Some people use calculators to balance their checkbook with their bank statement. Other people may use estimation to keep track of the activity in their account. The following activity lets students see how effective each method is and helps them decide which method would be more appropriate for them.

Monthly Statement

Date	Debits	Credits	Transaction Description	Balance
09/04		1500.00	opening deposit	1500.00
09/04	5.00		check printing fee	______
09/08	178.32		check #101	______
09/10	50.00		cash withdrawal	______
09/13	36.21		check #102	______
09/16	23.00		check #103	______
09/18	50.00		cash withdrawal	______
09/26		148.33	deposit	______
09/27	50.00		cash withdrawal	______
09/30		4.48	interest payment	______
09/30	3.56		monthly maintenance charge	______

1. Using the monthly statement above, have students estimate the balance in the account at the end of the month. They should do this mentally by following these steps:

   - Working with the debits column, group together amounts that add to approximately $100, and then add these groups together.

   - Estimate the total for the credits column. Finally, find the difference between these totals.

2. Have students use their calculator to find the exact final balance that should appear on the last line in the balance column.

3. Now have students compare their estimates from step 1 with the final figure they arrived at for the monthly statement ($1,256.72). Discuss the difference between the amounts and possible causes. Ask students, "Would your estimate be adequate if this were your checking account?"

Option: Some students may not have checking accounts. Go to local banks and savings and loan institutions to get brochures that describe the different accounts available and the costs involved. At the time of this writing, some banks in my area offer an interest-bearing account with no per-check charge and no minimum balance. It would be hard to find the disadvantages of such an account for anyone. Take time to explain the whole procedure, from writing the checks to balancing the checkbook with the monthly statement. You will be doing your students a great service.

LESSON 3

Equivalent Equations: Addition and Subtraction

Objectives

To further their understanding of the relationship between addition and subtraction, students will

- write equivalent equations;
- translate problem situations into mathematical equations;
- solve equations by rewriting them as equivalent equations.

Background

Using Equations

This lesson introduces the use of equations to solve problems. Equations contain equals signs and can be *solved* for the value of the unknown, whereas expressions can only be *evaluated*. By using variables in equations to mathematically state the situations, students can progress to more complicated problems. Students don't need to know how to solve a problem in advance, only how to mathematically describe a situation.

When we are working with variables, the step of finding the answer becomes one of solving the equation. At the level of the GED Test, equations can be solved by writing equivalent equations in which the variable is alone on its side of the equation. This method avoids the introduction of disconnected rules and also strengthens the students' understanding of the operations of addition and subtraction. It requires the use of common sense.

At its simplest level, solving equations can be used as a review of addition facts. By asking a student to solve the equation $3 + x = 10$, you are asking, "What number, added to 3, equals 10?" The student who does not want to admit needing to review the basic facts may enjoy this approach because he or she is "doing algebra."

Lesson Recommendations

Page **18**

Mental Math Exercises

This group of mental math exercises reviews translating expressions into mathematical operations. Students will then be prepared for working with equivalent equations later in this lesson.

Pages **18-19**

Writing Equivalent Equations

You can make the concept of equivalent equations less threatening to students by referring them to elementary applications that they will all recognize.

First, remind them that when they learned the addition facts, they also learned the subtraction facts. Illustrate with a simple example:

By learning $2 + 3 = 5$, they knew $5 - 2 = 3$ and $5 - 3 = 2$.

Second, they used this relationship when they checked subtraction answers by using addition.

Recall
$$\begin{array}{r} 100 \\ -\ 74 \\ \hline 26 \\ +\ 74 \\ \hline 100 \end{array}$$

In equation form:

$$100 - 74 = 26$$

Check:

$$26 + 74 = 100$$

Pages **20-21**

Writing Equations to Solve Problems

The words *solving word problems* have struck terror into many algebra students, perhaps even into some math teachers. Since we are limiting the word problems in the student book to practical, everyday situations, students should not experience the same reactions. Have students follow these steps:

1. Assign a variable. Have students use the first step to assign the unknown quantity a letter (variable). Without this step, the whole process loses its logic.
2. Reread the problem, looking for a sentence that describes the relationship between the known quantities and the unknown.
3. Place the equals sign. Each side of an equation has the same value. Look for places where you can substitute the verb *is* or *are*; these are likely places for the equals sign.
4. Translate the words into mathematical symbols. Compare this to translating between languages. Sometimes you can translate literally; other times you must rearrange the words. Often, rearranging into chronological order makes the process of translating easier.

Throughout this book we will explore many problem-solving strategies. However, there is no strategy that eliminates the need to *understand the problem.* If you suspect that your students are having difficulties because they don't understand what they are reading, then take some additional time to discuss the problem in other words. Talk about the situation from different perspectives and in students' own words before you begin trying to interpret it as a mathematics problem.

The two techniques in this lesson—following the action of the problem and using a known relationship as a guideline—will help the student who does not have the insight and experience to be able to analyze a situation. Both techniques provide help in writing the original equation. Once that is done, students can let the mathematics take over. Math manipulations will dictate the path to the solution. After the students find the solution, they must go back to the original word problem and see if the answer makes any sense as a solution. Is the answer reasonable?

English Problem Situation —interpret→ Mathematics —manipulations→ Solution —reasonable?→ Original Problem Situation

Pages **22-23**

Solving Equations

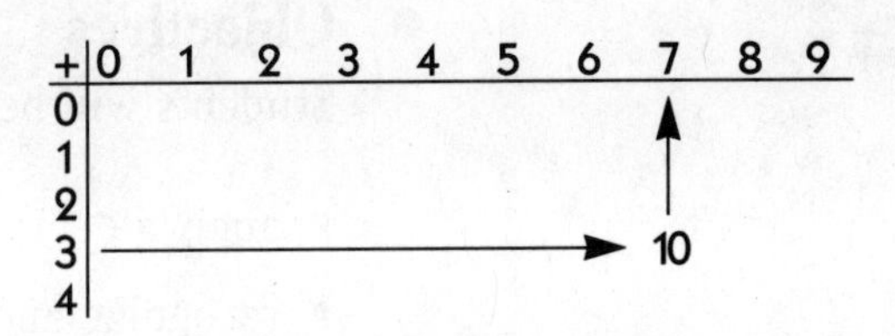

Problems of the form $3 + x = 10$ are most likely to occur in applications. They represent the "missing addend" interpretation of subtraction. You can demonstrate an elementary version of this by using the addition table found on page 101 of this guide. For example, to explain the problem $3 + x = 10$, locate the 3 in the row headings and move sideways into the table entries until you come to the 10. Then follow that column back up to its heading to find the value for x (7).

The student should have some experience with these manipulations before moving on to using this skill in solving problems.

You are teaching students to use common sense to manipulate equations, rather than to follow the transformation rules of algebra. You may think that this is harder than simply following the rule to add or subtract the same thing to both sides of the equation. The rules work, of course, but they are often so abstract that most students at this level fail to retain an intuitive understanding of what an equation and its solution are.

A technique to use if a student "forgets" even the commonsense manipulations required to solve a problem is to *compare a complicated equation with a simple arithmetic fact.*

For example, compare $x - 9.6 = 2.4$ to $10 - 7 = 3$.

Ask how the fact would be rewritten so that 10 is alone: $10 = 7 + 3$

This gives students a concrete example of how to proceed with the more complicated equation:

$$x = 9.6 + 2.4$$

Page **24**

Another Way to Subtract

Help your students understand this section better by using a number line as a visual. Locate the two numbers 26 and 9 on a number line. Then move one place to the right from both numbers to 27 and 10. Note that the difference (or interval between the numbers) remains the same for both problems.

LESSON 4 • Geometry Topics

Objectives

Students will be able to

- apply algebraic problem-solving techniques to geometry topics;
- recognize and measure right, acute, obtuse, and straight angles;
- use the facts about complementary and supplementary angles to find a missing angle measure;
- use the fact that angles of a triangle total 180° to find the missing measure of one angle;
- recognize the equality of vertical and corresponding angles;
- apply knowledge of angle equality and reasoning skills to complex figures.

Background

Geometry and the GED Test

The geometry problems that appear on the GED Test do not involve difficult concepts. Many can be solved using just addition and subtraction if students know the vocabulary. This lesson explains the meaning of the terms and gives students the opportunity to solidify the techniques of the previous lesson. Make this a hands-on lesson using real objects when you can.

Lesson Recommendations

Page **28**

Mental Math Exercises

Apply the compensation techniques discussed in the last lesson to these problems, which are set up as equations. Guessing is an appropriate means of solution.

Pages **28-29**

Angles

The measure of an angle is the "amount of turning" that takes place between its two sides. Demonstrate this on the board or overhead projector by placing the point of a pencil on the vertex of an angle while the pencil lies along one side of the angle. Then rotate it, with the point fixed on the vertex, to the other side. Make this idea real to students by reminding them of the everyday uses this concept has in our language. "He did a 180" describes a person who changed his mind to the other point of view. "She came full circle" describes a person who came back to her original view.

Use a protractor to measure the angles pictured in the discussion of right, straight, obtuse, and acute angles. Verify that the measure of the acute angle falls between 0° and 90°.

EXTENSION ACTIVITY: **ESTIMATING ANGLES**

To develop estimation skills with angles, have students construct an estimating protractor. You may want to draw a sketch of each step on the board or have students follow along as you do the activity yourself.

1. Fold a piece of paper precisely in half, taking care to line up the edges.

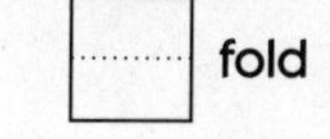

2. Fold the paper again, aligning the edges with the previous fold.
3. Start cutting on the closed end of the paper, and cut a curved line to the opposite bottom edge.

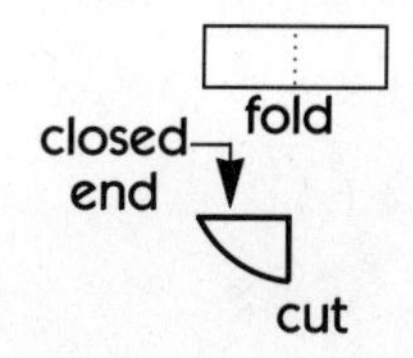

4. Fold the paper in half once more, forming a "pie wedge."
5. Unfold the paper twice, so half a circle is showing, and label the fold lines with the corresponding degrees.

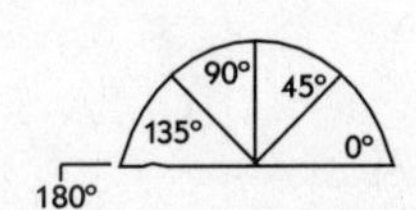

Ask the students, "How would you label the protractor so you could measure in both directions as with a standard protractor?"

6. Add the second set of numbers to the protractor.

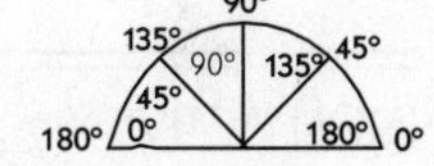

Discuss how students would know which set of numbers to use. Only one of the measures *makes sense* for the angle being measured. The angles in **Problem 1** can be measured using these "estimating protractors."

Students should also be able to judge an angle's approximate size (within 30°) just by looking at it.

Pages **30–31**

Complementary and Supplementary Angles

Students should try to learn the meanings of the words in this lesson, and they must be able to use them in context. For the GED Test, students must know that a right angle measures 90°, a straight angle measures 180°, and a full circle contains 360°. Most of these terms (except *right angle*) will be explained by their use on the GED Test.

Page **32**

Triangles

A key concept on the GED Test is that the sum of the angles of a triangle equals 180°. If students cannot immediately see the path to the solution, they may use a formula as a guideline (for example, $\angle a + \angle b + \angle c = 180°$). It is a standard and reliable technique. But many adults in your classes will have a highly developed reasoning ability. For these students it is almost painful to write out the steps as they are in the book. It is OK to do some of the steps mentally. However, insist that everyone write the problem, whatever form that takes. Students need to be reminded that often it is the expression that is desired on the GED Test and not the final answer. This skill is a critical one in preparing to take the GED Test.

Continue to use the steps, not only for the students who need them, but also so the others can compare their equations to the ones given. Some very important class discussions will result from trying to decide whether the equations are equivalent.

A very important concept is illustrated, but not noted, in the solutions of the examples. The minus sign preceding the parentheses "distributes" to both addends. You may also want to take time to work the second solution as shown below.

$$180 - (57 + 99) = b$$
$$180 - 57 - 99 = b$$

Subtracting from left to right: $180 - 57 = 123$
$123 - 99 = b$
$24 = b$

Students need to compare the procedure above with the procedure where the numbers in parentheses are added first and then their sum is subtracted. **Problem 9** tests their ability to discern the difference between expressions like these.

EXTENSION ACTIVITY: **ADDING ANGLES**

Do this extension activity only after students are sure that a straight angle contains 180°. This activity reinforces that principle.

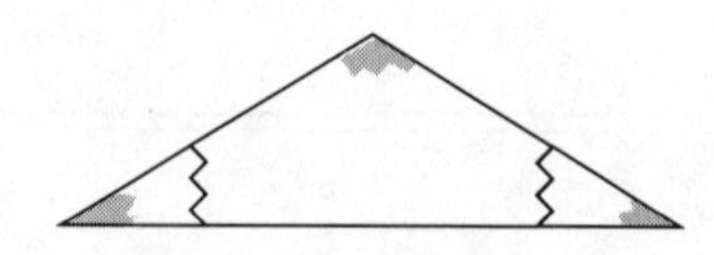

1. Cut out a variety of different-shaped triangles, then give each student a triangle.
2. Have students mark or color each corner of their triangle.
3. Instruct them to tear each triangle into three pieces, each containing a corner.
4. Have them line the pieces up so that the original corner angles abut each other on a straight line, perhaps the edge of the desk. In every case, the sum of the three angles will make a straight line.

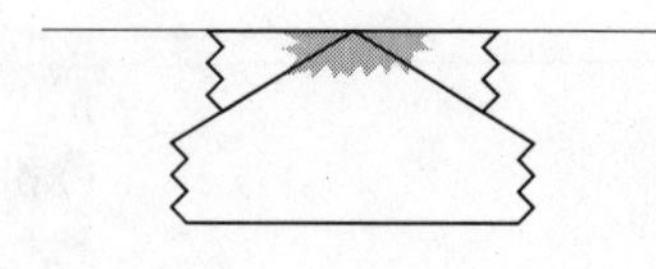

Pages **33-34**

Special Types of Angles

The equality of corresponding angles depends upon the lines being parallel. When taking a test, students should not assume lines are parallel or perpendicular just because they look that way. That fact must be stated explicitly or be deducible from the given information. The GED Test does not try to trick examinees, but being careful about this is important in other situations.

LESSON 5 • The Number Line and the Coordinate Graph

Objectives

Students will be able to

- recognize the order relationship with negative numbers;
- add and subtract on the number line, using negative as well as positive numbers;
- name the position of points on the coordinate graph, using Cartesian coordinates.

Background

Working with Integers

In this lesson, the operations of addition and subtraction are extended to all the integers, positive and negative and zero. This preparation course attempts to find relevant applications for algebra topics. Your students need concrete representations of the concept of negative numbers, not abstract rules defining their use.

As a natural extension of the number line, graphing on the coordinate plane also is introduced. Explain to students that the axes are two number lines that intersect perpendicularly. The positioning of points in all four quadrants is stressed here, and ordered pairs (x, y) are used as a way to locate points on a plane.

Lesson Recommendations

Page **36**

Mental Math Exercises

Use the mental math exercises to show students that by estimating, they know a lot about a problem. Point out to students that if they use $20 + 13$ as an estimate for $18 + 13$ in Exercise 1, for example, they also know that the exact answer is less than their estimate. Later in the book, the notation 33^{-} will be used to indicate this result.

Page **36**

The Number Line

Use the diagram to introduce the number line and the presence of negative numbers. Discuss the symmetry centered at zero. Remind students that anything to the left of a value on the number line is less than that value. Discuss the meaning of the arrows at both ends of the number line.

Pages **37-38**

Adding and Subtracting on the Number Line

Once again, try to avoid using the algebraic rules that detract from a student's intuitive grasp of these concepts. Your students should have no trouble with the simple examples given, even though some of the notation is foreign to them.

Another application that you can use is to apply these concepts to the checking account activity (Calculating and Estimating with a Checking Account) on page 14 of this guide. Depositing money is the addition of a positive, and writing a check can be interpreted as either subtracting a positive or adding a negative (assuming you are dealing with a bank that will cover bad checks).

EXTENSION ACTIVITY: **VISUALIZING CHECKING ACCOUNT ACTIVITY**

Handout 3
Number Lines
TG Page 103

Using the checking account activity on page 14, have students record the transactions on an appropriately labeled number line. It is important for students to visualize a new concept whenever possible. By taking time to locate the numbers, count the spaces, and discover the landing spot, you will help students at all levels to grasp the ideas. A full page of number lines is included for your convenience on page 103 of this guide. Make copies for your students as well.

Page **39**

Using the Number Line

Note that no mention was made of subtracting a negative. Since this subject often serves as a source of confusion for students, it is introduced only in context, as in **Problem 4b**. This problem asks for the difference between the recorded and the windchill temperatures. The strict interpretation of this problem would be the expression $12 - (-10)$. But to a student looking at the thermometer, it is obvious that the problem calls for adding the absolute values of the two numbers.

Your class may be able to understand the strict interpretation and its method of solution. But most classes will just accept this as a question about the positions of the two numbers. Since questions of this type will be on the test only in a real-life context, your students will not be impaired by this interpretation.

Few of the problems in this lesson will tempt students to use their calculator. However, this is a good opportunity to introduce the use of the [+/−] key. You can enter a number and then change its sign by pressing this key. For example, **Problem 5e** would be entered into the calculator in this order:

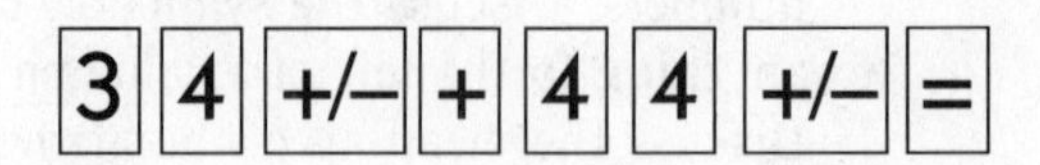

Have students use their calculator to check the other problems in this section.

Pages **40-42**

The Coordinate Graph

To introduce the concept of locating points by means of coordinates, you could use a map and locate points of interest by using the locator. (For example, one map shows Austin, Texas, at E-27 on the map of Texas. These coordinates will vary depending on which map you use.) This analogy is not strictly parallel to a coordinate graph, because on a map, the spaces are labeled, and on a Cartesian plane, the lines are labeled. But, this method will still help prepare students to use two values to find a point.

Introduce the Cartesian plane by studying the diagram. Note the symmetry about zero. Discuss the four quadrants that result from the intersection of the axes. Stressing that the starting point is always the origin (0, 0), begin locating points in the first quadrant. Ask students: "Where are the negatives? How would you get to a point in the second quadrant?" The region to the left of the origin is negative just as it was on the number line. Characterize the points in each of the quadrants as indicated by the following sketch.

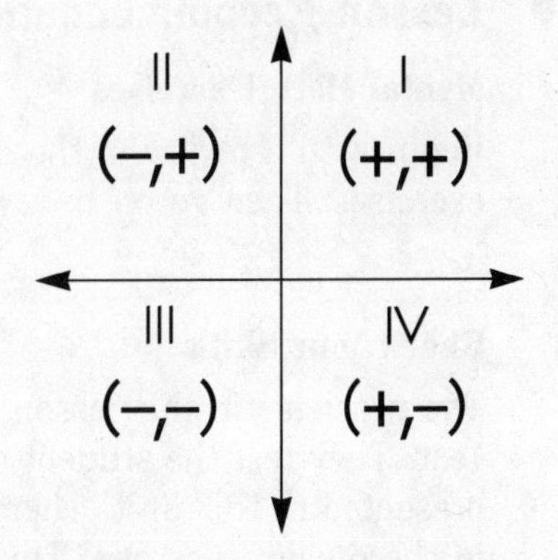

Finally, discuss the points that lie *on* the axes. Reproduce the grid on page 42 (of the student text) as a transparency, and ask the students to locate the following points:

(3, 2) (2, 3) (6, 3) (3, 6) (1, 5) (−1, 5)
(5, 1) (−5, 1) (−1, −5) (−5, −1) (5, −1)
(0, 3) (3, 0) (−3, 0) (0, −3)

The coordinates of a point make up an **ordered pair**. This name is an indication of the importance of the order of the two values in the set. The above examples, as well as the problems on the assignment page (page 42 of the student text), will reinforce the importance of order for students. The idea of graphing ordered pairs appears later in **Lesson 24**.

LESSON 6 • Checkpoint I

Objectives

Students will

- review the methods covered in this unit;
- evaluate their progress using the type of problems that will be on the GED Test.

Lesson Recommendations

Page **44**

Mental Math Exercises

Begin with a review of the mental techniques represented in the mental math exercises. Then go on to review the main ideas of this section.

Pages **48–49**

Check Your Skills

The problems in this lesson are typical of the problems on the actual GED Test. They test the student only on the addition and subtraction concepts presented in this unit. There are some problem settings that are not discussed in the previous lessons. This is to stimulate students to apply what they have learned to a new, but similar, situation—the test of real learning in terms of problem solving. To simulate actual test conditions, the students should have no more than 15 minutes to complete these problems.

After the problems have been checked, there should be a full discussion of every problem. The individual students' methods of solution should be explored. You should look for and point out the good aspects of their reasoning, even if errors were made, so students will still be encouraged. Discuss the fact that they would not have used a calculator to find the answer for these problems.

Problem 1: This problem tests the student's understanding of size relationships with decimals.

Answer Choices:

(1) Nearly correct. The error was in not recognizing 1.2 as equivalent to 1.20.

(2) and **(3)** More serious errors involved ignoring the significance of the decimal point.

(4) Correct.

(5) Nearly correct. The error was in thinking that the number of places behind the decimal point determines the value.

Problem 2: This is a set-up problem and requires knowledge of equivalent expressions.

Answer Choices:

(1) Starts with balance, subtracts both the checks and the deposits.

(2) Correct.

(3) Starts with balance, adds both the checks and the deposits.

(4) This choice would be correct if the balance were added instead of subtracted at the end.

(5) Subtracts the balance and adds the checks; shows confusion about the situation. Students who choose this answer could use more time on the checking account exercise (TG page 14).

Problem 3: This problem applies the concept of addition by grouping. Have students note that only the correct answer came close to making sense in the situation.

Answer Choices:

(1) Added two groups, but then subtracted.

(2) Correct.

(3) Doubled the result. Choosing this answer could go back to a misunderstanding of the formula, $2l + 2w = p$.

(4) Multiplied 25×35. Someone who chose this may have been thinking of the area formula.

(5) Students would have chosen this only if they didn't know what else to do.

Problem 4: This is a set-up problem that requires subtraction and knowledge of variables.

Answer Choices:

(1) Multiplied the present weight by the amount of weight to be lost.

(2) Added the amount to be lost by the present weight.

(3) Added the present weight to the amount to be lost.

(4) Correct.

(5) The order is incorrect. Refer students back to "original number comes first," explained on page 4 of the student book.

Problem 5: Even though they are not in evidence, this problem tests the concept of negative numbers. Advise the students to sketch this situation to discover their errors.

Answer Choices:

(1) Doubled 37, then subtracted.

(2) Subtracted the quantities. After all, it is a comparison!

(3) Correct.

(4) Doubled 37, then added.

(5) Students were confused by the problem.

Problem 6: This set-up problem requires experience with the concepts of discount and rebate.

The errors made in each response are obvious. Those who added may have responded to the word *total* in the problem. Be sure to stress looking for relationships, not clue words. Correct answer: **(2)** \$9,782 − \$7,990.

Problem 7: This problem requires prior knowledge that a straight angle measures 180°. It would be an interesting exercise for the class to make up 5 possible answer choices if this had been a set-up problem.

Answer Choices:

(1) and **(2)** As here, students often choose answers that contain the numbers in the problem.

(3) Correct.

(4) In this case, the two numbers were added, but the second step of subtracting from 180° wasn't completed.

(5) This answer indicates that the student didn't understand the problem.

Problem 8: This is a problem-solving situation requiring the prior knowledge that $\angle a + \angle b + \angle c = 180°$.

Answer Choices:

(1) and **(2)** These answers use numbers directly from the problem, or students could have misread the question.

(3) Correct.

(4) Here students began correctly, but did not subtract the sum from 180°.

(5) Students may not have recognized that, because the two angles are equal, there is enough information.

EXTENSION ACTIVITY: **COMPUTATIONAL CHECKS**

This is also the time to check for the student's proficiency with the addition and subtraction single-digit facts. Some suggestions follow to make this more fun than the traditional methods for using flash cards.

Suggestion 1: Pair the students. After stating the problem, one student drops a ball held at shoulder height. The other student should give the correct response before the ball hits the ground.

Suggestion 2: Group the students, and have at least one student in each group who can serve as the leader. Remove the face cards from a deck of playing cards, letting the aces equal one. After shuffling the deck, the leader "plays" two cards. The students take turns adding the two numbers showing on the cards and responding with the correct answer.

LESSON 7 • "Seeing" Multiplication and Division

Objectives

Students will be able to

- develop intuition for deciding from the context of real-life problems when to multiply or divide;
- write and understand multiplication and division problems in a variety of formats;
- use variables in setting up problems;
- translate verbal problems into mathematical notation;
- use knowledge of basic facts to develop techniques for mental computation and estimation.

Background

Multiplication and Division Facts

Quick recall of the multiplication facts of numbers 1 through 10 is essential for success with math. Students will also benefit from knowing the facts for 11 and 12.

Work individually with students who are struggling with the basic combinations. As you did with addition, help the student construct a personal multiplication table like the one on page 101 of this guide (see TG pages 6-7 for instructions). The student can then use the table when needed.

1. Have the student fill in the answers he or she already knows. Remind the student that multiplying by two is the same as doubling the number.

2. As you proceed, point out the patterns that occur.

a) For example, note the symmetry about the diagonal of perfect squares. (If you fold the paper on that diagonal, the numbers that fall on each other will be equal.)

b) You could also build on the example of adding eights from the first lesson. Find each successive multiple of eight by adding eight to the previous one ($8 \times 3 = 24$: $24 + 8 = 32$, so 8×4 must be 32).

Many people have the most difficulty with the 16 answers in the bottom right square of the multiplication table. This group can be made less intimidating by the following tips:

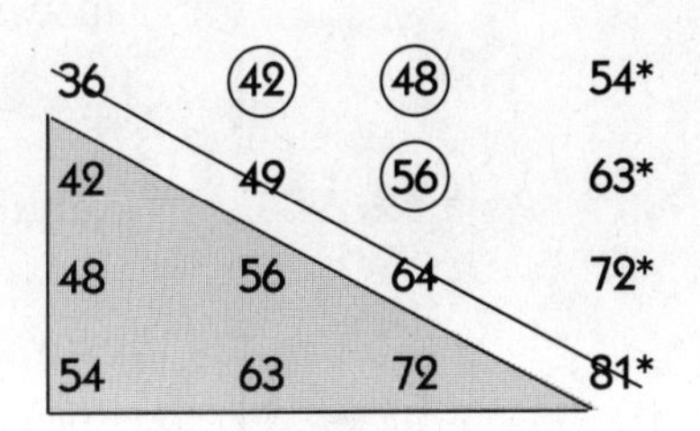

1. Diagonally divide the square containing these 16 products into two symmetric halves. The numbers on the line (perfect squares) are critical ones to know without hesitation.
2. The shaded products are the same as the upper half, so there are really only 10 facts to learn, instead of 16.
3. The starred (*) numbers are multiples of 9. Point out that the digits of each multiple of 9 add up to 9 or a multiple of 9.
4. That leaves only three more facts to memorize—the worst for most people being $7 \times 8 = 56$. Use this fact as often as you can in your classroom examples. The student book also uses this tactic of constantly barraging the student with the facts that are troublesome.

• Lesson Recommendations

Page **50**

Mental Math Exercises

The purpose of these mental math exercises is to lay the groundwork that will connect multiple addition to multiplication.

Page **50**

1. Picture the Situation

As with addition and subtraction, the emphasis here is on the relationships of *combining equal* groups for multiplication and *separating* into *equal* groups for division.

As a second interpretation, demonstrate multiplication as repeated addition and division as repeated subtraction. **Example C** on page 50 pictures this, as it shows division as an operation that determines the number of equal-sized groups in a number. (How many times can you subtract 8?) **Example D**, in contrast, uses division to determine the size of a given number of groups. It does not lend itself to the interpretation of division as repeated subtraction.

Page **51**

2. Write the Problem

Use **Problem 4** to emphasize that addition and multiplication are commutative, whereas order does make a difference in subtraction and division.

While doing **Problems 4c** and **g**, your students may note that the answers given on the calculator differ only slightly: one answer has a negative sign in front, while the other answer doesn't. The following fact will always be true:

$$(a - b) = -(b - a)$$

However, you may need to remind students that 22 does not equal −22.

Note: Whenever you read a division problem aloud, stress the preposition *by,* as in "45 divided *by* 9." It is also helpful to restate the division problem by asking, "How many 9s are there in 45?" This interpretation will help the answers make sense when students are later dividing by numbers less than 1.

Page **52**

Notation

Using the form $\frac{45}{9}$ to indicate "45 divided *by* 9" may be strange for some of your students. Be prepared for students to respond (with some anxiety), "But these are fractions!" You should agree, but explain that the fraction bar is also a way to represent division.

Compare and discuss the placement of the numbers in $\frac{45}{9}$ and $9\overline{)45}$. Students may find it easier to read (top to bottom), "45 divided by 9." The pedagogical advantages of this fraction notation will become apparent as your students move effortlessly into the study of fractions in later chapters.

Page **53**

Translating Words to Problems

Note that addition and subtraction problems also are included in the examples of mathematical expressions. You should point out the similarities (e.g., addition and multiplication are both combining operations), as well as the differences (e.g., multiplication requires equal groups) across the operations.

Problem 7 is a challenging one. Your students will benefit from some sketches or actual objects to manipulate to decide whether the words ask them to combine or separate.

EXTENSION ACTIVITY: **EVERYDAY MULTIPLICATION AND DIVISION**

After you have isolated some of the multiplication and division facts that are troublesome to your group, choose one fact, say, $6 \times 8 = 48$. As a homework assignment, ask the students to find a situation in their own lives that can be represented by this problem. One student may notice that there are 48 sodas in 8 six-packs; another may earn \$6 per hour and work 8 hours a day, earning \$48 per day. This activity reverses the book's procedure of asking students to write the problem that describes the situation.

Page **54**

3. Find the Answer—Mentally

Using an overhead projector, show a multiplication table you constructed with a student or the one provided on page 101, and point out that the first row and column illustrate the principle of 0 in multiplication. Also, show that the second row and column are identical to the row and column headings—1 is the **identity element** for multiplication.

Both of these properties are more complicated for division because of the difficulties involved with the order of numbers. The examples point out the importance of being careful.

Page **55**

Building on Basic Facts

The estimation techniques introduced in this lesson are just the beginning steps of what will become a powerful way of thinking for your students. The students will probably not feel comfortable during this introductory phase, but by being persistent, you can make them feel "at home" with numbers. Building number sense is a major goal of this method of instruction.

Pages **56-58**

Rounding and Estimating

When rounding, students automatically want to round to the nearest hundred or thousand. This is all their previous training taught them to do. For this reason, it may be difficult to get students to round to a compatible number for division.

Parts **a**, **b**, and **c** in **Problem 13** provide good practice for finding compatible numbers for division. In **13a**, for example, have students look at the denominator first. Ask students, "Will 7 go into 4? Will 7 go into 41? In this problem, what number would you choose to divide 7 into?" Students should choose 42. The divisor determines how many digits are needed in the numerator to divide. The rest of the digits in the numerator can be regarded as zeros and placed in the answer.

The point here is to change students' focus for easier estimates. Students should work from left to right to find a compatible number to divide into. The next set of problems explains and reinforces compatible numbers.

The mental computation problems depend on the students' knowledge of the multiplication tables. Of course, students will review them as they do the problems, but using the following extension activity as a nonthreatening review would be a good idea before you begin **Problems 13**, **14**, and **15**.

EXTENSION ACTIVITY:

SIEVE OF ERATOSTHENES

Handout 2
Hundreds Chart
TG Page 102

Before you begin this activity, explain **prime numbers** (a number that can be evenly divided by only itself and 1) and **composite numbers** (a number that can be divided equally by other numbers as well). This activity will give students the most effective display of patterns formed by **multiples**.

Give each student a copy of the hundreds chart (page 102 of this guide). Then, instruct them to follow the steps below to see the patterns that will be formed.

1	2	3	4	5	6	7	8	9	10
11	12	13	14	15	16	17	18	19	20
21	22	23	24	25	26	27	28	29	30
31	32	33	34	35	36	37	38	39	40
41	42	43	44	45	46	47	48	49	50
51	52	53	54	55	56	57	58	59	60
61	62	63	64	65	66	67	68	69	70
71	72	73	74	75	76	77	78	79	80
81	82	83	84	85	86	87	88	89	90
91	92	93	94	95	96	97	98	99	100

1. Look at the first row. Cross out the multiples, and put boxes around the prime numbers up to 9. By the time you get to 9, you will want to be more efficient.
2. With a black marker, cross out the multiples of 9. Note that they form a diagonal across the page and that the two digits in each of those numbers add up to nine (or a multiple of 9, as in 99).

3. How can you detect a number that is divisible by 5? With a red marker, draw a line through the multiples of 5 in the columns with numbers ending in 5 and 0.

4. Is there an easy way to recognize a multiple of 2? With a green marker, draw a line through all the columns whose last digit is even. The only remaining prime numbers in the first row are 3 and 7.

1	2	3	4	5	6	7	8	9	10
11	12	13	14	15	16	17	18	19	20
21	22	23	24	25	26	27	28	29	30
31	32	33	34	35	36	37	38	39	40
41	42	43	44	45	46	47	48	49	50
51	52	53	54	55	56	57	58	59	60
61	62	63	64	65	66	67	68	69	70
71	72	73	74	75	76	77	78	79	80
81	82	83	84	85	86	87	88	89	90
91	92	93	94	95	96	97	98	99	100

5. Start with 3. Put black *X*s on all the multiples of 3. Six is even, 9 is odd, 12 is even, 15 is odd, etc. When you are finished, you will note that these also form diagonal patterns. They also show that multiples of 3 are every third number. Why don't you have to cross out any multiples of 4? (They were all eliminated when you crossed out the multiples of 2.) Are there any multiples of 6 that are not crossed out? (No, the combination of multiples of 2 and 3 took care of all of them.)

1	2	3	4	5	6	7	8	9	10
11	12	13	14	15	16	17	18	19	20
21	22	23	24	25	26	27	28	29	30
31	32	33	34	35	36	37	38	39	40
41	42	43	44	45	46	47	48	49	50
51	52	53	54	55	56	57	58	59	60
61	62	63	64	65	66	67	68	69	70
71	72	73	74	75	76	77	78	79	80
81	82	83	84	85	86	87	88	89	90
91	92	93	94	95	96	97	98	99	100

6. What is the first multiple of 7 that is not yet crossed out? (It is 49 because smaller multiples were crossed out when you did the smaller primes. For example, $35 = 5 \times 7$ was included in the multiples of 5.) You will also need to cross out 77 and 91. Using this same logic, is it possible for any of the unmarked numbers to be multiples? (No, the next prime number is 11, and $11 \times 11 = 121$.)

Take time to list all the prime numbers less than one hundred. Point out once more that a prime is a number that is not divisible by any number other than itself and 1. Familiarity with these will increase students' number sense.

LESSON 8

Measurement: Multiplying More than Two Numbers

Objectives

Students will be able to

- use units of measure in both the English and metric systems;
- recognize which attribute (length, area, or volume) is represented in a problem, so they can choose the proper formula;
- recognize the shapes of a rectangle, parallelogram, and triangle, so that they can select the method or formula for finding area and perimeter;
- multiply with more than two multipliers, using grouping of compatible factors to make it easier to mentally find the answer.

Background

A Hands-On Experience

This is a hands-on lesson. Do not allow this to become merely a lesson in finding formulas and substituting values in them. Experience shows that students need to manipulate the materials in order to make the applications of this subject relevant. The following materials and equipment are necessary to make this lesson real for students:

- Measuring tapes and rulers for English and metric systems.
- Centimeter and/or 4-to-the-inch grids on transparencies (TG page 108).
- Cubic centimeter blocks and cubic inch blocks.
- Envelopes containing precut shapes. Include triangles (two of which are identical), rectangles, parallelograms, and other polygons of four, five, and six sides.

Lesson Recommendations

Page 60

Mental Math Exercises

These mental math exercises provide a review of the important rules for zero and one. These rules are fundamental and should be mastered by students before continuing.

Page 60

Units of Measure

Each student should have the experience of measuring with both the English and metric systems. Using the measuring tapes from both systems, expand **Problem 1** so that the students also measure their height, hand span, and arm span using both measures. An additional benefit of this exercise is that students will become aware of some personal references of measure that will always be available to them. For example, a yard or meter is often the distance from a person's nose to the fingertips of his or her outstretched arm, an inch may be the length of the last joint of one's index finger, and elbow to wrist may be a foot in length. Be certain that each student goes away from this exercise with at least one of these "personal measures."

Expand **Problem 2** by asking which metric unit would be used for each measurement. Answers: **a)** km, **b)** m, **c)** m^2, and **d)** m^3.

Page **61**

Areas

Many people who take the GED Test make the mistake of using the perimeter formula for area and vice versa. This error indicates a true lack of understanding concerning the attributes of length, perimeter, area, and volume. Some students with learning difficulties will not understand what distinguishes one from the other unless they actually "feel" area by rubbing their hands over something like the desktop and "feel" perimeter by running their fingers along the sharp edge around the desk. (Thanks to Mary Jane Schmitt of Massachusetts for this observation.)

Even though some of your students already know the formulas for area that are in this lesson, have them participate in the discovery approach to finding area. What is discovery for some will be reinforcement for others. The discovery of the easiest way to find the number of square units in any rectangle should not be difficult for students. In fact, when they determined the number of "palms" in the surface of their desk (page 61), many naturally found the number in each row and the number of rows that could fit.

Page **61**

Rectangles

It is interesting to note that in the real-life situation described in **Problem 5**, the estimate of square footage is the one that gives some allowance for waste and would be more useful in buying tile. You might want to mention that some tile layers like to add a 10% waste allowance to their measurement.

EXTENSION ACTIVITY: WORKING WITH RECTANGLES

Give small groups of students envelopes of the precut shapes listed on page 33 of this guide and a transparent grid. (Either the centimeter or 4-to-the-inch grid will work, but have your shapes cut to fit the one you choose.) Instruct students to take the **rectangles** out of the envelope. Ask them, "What is it about these shapes that makes them rectangular?" (Important characteristics to cover are number of sides (4), opposite sides that are equal and parallel, and right angles.)

EXTENSION ACTIVITY: **HANDS-ON MEASURING**

Bring several boxes of various sizes (cereal, medicine, etc.) to class and do some hands-on measuring. Show students measurements of *length* on the different boxes. Find *regions*, and help students to figure the *area* of the sides, top, and bottom for each box. After students have completed page 65, help them find the *volume* of each box.

Page **62**

Parallelograms

By following the procedure to visualize a parallelogram as a rectangle, use the cutout shapes and transparent grids so your students can see that the formula for the area of parallelograms makes sense. Allow them to actually cut the parallelograms and rearrange the pieces to form rectangles.

It may be difficult for students to recognize that the height of a parallelogram is different from the length of the slanted side. Point out that this height is perpendicular to the side that they determine to be the base (not necessarily the bottom side) and may be measured outside the figure. (See **Problem 7b**.)

Pages **63-64**

Triangles

Have students use two identical triangles from the envelope and a transparent grid to visualize the connection between the area formulas for triangles and parallelograms.

Pages **64-65**

Irregular Figures

Irregular figures are very likely to be represented on the GED Test, since they present a problem-solving situation as well as a test of geometry skills. Review with students the strategy of dividing the complex figure into smaller shapes for which they know how to find the area. The first example shows the addition of the one smaller area, while the second example shows that one area can also be subtracted to find the remaining area.

Page **65**

Volumes

Finding the volume of a rectangular container follows the same logical pattern as finding the area of a rectangle. It would be most instructive if you had a small box that you could fill exactly with layers of actual cubic inches or cubic centimeters. Some of these instructional materials are available at parent/ teacher stores or in instructional materials catalogs. They also can easily be constructed by a cooperative industrial arts class.

Page **66**

Finding the Answer Mentally

The section on compatible numbers expands on the associative and commutative properties discussed earlier. Compatible factors are two numbers that can be multiplied mentally to get a "nice" number (one that is easy to work with). For example, any multiple of 5 can be multiplied mentally by any multiple of 2. This can be expanded to the special cases of 4 and 25, and 2 and 50. These mental gymnastics are not difficult to carry out when students understand the underlying principles. Encourage them to write down the factors as suggested in the examples so that they don't lose track of them.

The section on doubling and halving is included here because of the formula for the area of a triangle. The technique of breaking the number up into its parts and then multiplying or dividing each part by 2 is introduced here almost intuitively. **Lesson 10** will review this technique, using the distributive property as reinforcement.

Your students should be gaining a feeling of empowerment as they learn these techniques and become more comfortable with numbers. Encourage your students to do most of the problems in the Check Your Understanding set by using mental techniques rather than paper and pencil or a calculator.

LESSON 9

Equivalent Equations: Multiplication and Division

Objectives

Using two of the most practical relationships in daily life, $d = rt$ and $c = nr$, students will

- explore all aspects of the relationships $d = rt$ and $c = nr$;
- visualize multiplication relationships, using both a table and a graph;
- learn to write equivalent equations using multiplication and division;
- solve problems by writing equivalent equations;
- develop number sense about what happens when they multiply and divide.

Background

Using Rates

This lesson provides the foundation for the study of rates. In later lessons, this will lead to ratios, proportions, percents, and even fractions. A thorough understanding of the basic relationship is critical to further success. For this reason, the scope of this lesson is limited to analyzing how the inverse operations of multiplication and division are related in these practical applications. The student text provides the structure.

Lesson Recommendations

Page **68**

Mental Math Exercises

These exercises review the mental math techniques from the previous lesson.

Page **68**

Finding Total Cost

The two examples at the beginning of the lesson provide motivation, as well as a reference to the concrete, which is vital for true learning.

First, analyze the table about CDs under the first example. Discuss both columns and the relationship between them. (As the number of CDs increases by 1, the cost increases by $14.)

Problem 1a requires students simply to supply a missing value. They could either add 14 to 56 or multiply 14 by 5 to find the cost of 5 CDs. This serves as a reminder that multiplication is repeated addition.

Problems 1b and **1c** lead students to generalize to discover the formula for unit price. They have to multiply the number of CDs (n) by the price per item (r for rate).

Finally, **Problem 1d** requires students to apply the formula. How much has Paul spent?

Taken one step at a time using a familiar situation, the mathematics involved is not difficult.

The graph on page 68 is included so that students can visualize what happens when they multiply with numbers greater than 1. As the number of CDs increases, the total cost increases. This is not news to anyone in the class. But it needs to be discussed, and the link between the concrete and the abstract has to be made. Make this generalization in order to introduce students to this kind of mathematical thinking.

Use this opportunity to familiarize students with this type of line graph. The questions will guide you. For **Problem 2a**, find the 8 and move straight up until you meet the graph of the line. From that point, move to the vertical axis to find the corresponding value for c (total cost). **Problems 2b-d** move the discussion into discovering the relationship between multiplication and division.

For example, in **Problem 2b**, students are given the total cost. By finding that value on the vertical axis ($100), they can determine the approximate corresponding value on the horizontal axis (7). Then they can analyze what has actually happened: they divided $100 by 14 to find that value. Apply that principle to some value not on the graph, then generalize: to find the number of items, divide the total cost by the price per item. If your class needs simpler examples, try talking about apples that cost 10¢ each and how many you could buy for 50¢.

Page **69**

Finding Distance

The second example, which analyzes the relationship $d = rt$, follows the same structure as the total cost formula. Your class should be able to move through it with the ease that comes from familiarity.

Pages **70–71**

Equivalent Equations

The preceding student pages intuitively explored the relationship between multiplication and division by relying on students' experience with concrete examples. This section formalizes that relationship analytically by showing the relationship between equivalent multiplication and division equations. After practicing equation manipulation (using numbers and then variables), students will apply what they have learned to the two formulas discussed earlier. Finally, they are asked to apply these techniques to problem situations.

Two different ways to approach formula problems are outlined in the examples on page 71. The first way substitutes known values into the basic formula. An equivalent equation is then written so that the unknown value is alone. The other approach uses students' knowledge of what the problem calls for. If students know they have to divide cost by rate, they're already using an equivalent form of the formula. Encourage these students to write the problems in equation format.

It may be helpful to your students to point out the following: $\frac{32}{8} = 4$. The 4 and the 8 multiply to equal 32. The 4 and 8 can be interchanged to form an equivalent fraction: $\frac{32}{4} = 8$.

The presence of a "not enough information" problem in the discussion set will cause some confusion. Do not allow students to get into a rut about problem solving, thinking that all the problems in a set will follow the pattern of the examples above it. **Problem 9** is strategically placed to alert students because there is also a similar problem in the individual exercises that follow the lesson.

Pages **72-73**

Analyzing Answers

The purpose of these class exercises is to make students aware of the range of answers before they multiply or divide. This will expand their number sense and allow them to make estimates and detect unreasonable answers that may appear on a calculator display. Again, the discoveries (that you will pay more than $12 for 12 gallons of gas when the price per gallon is over a dollar) will not be startling to students. Build on their everyday knowledge, and point out the underlying mathematical principle. *Make your students more aware of the mathematics they already know.* You are encouraging them to use this awareness more frequently and to use it to predict the size of answers before they multiply or divide.

Problems 13d and **17e** represent the main issues. You can give these more emphasis by providing further examples. The answers to **Problems 15** and **19** can vary, but the listed answers correspond to the principles discussed in this lesson.

Note: When discussing page 72 with teachers, I found that many "invented" ways to complete the table without using a calculator. You might challenge some of your more advanced students to find the cost corresponding to .85 and .99 by mental computation.

EXTENSION ACTIVITY: **DIVIDING BY SMALLER AND SMALLER NUMBERS**

Construct a table on the board that shows what happens when you divide 20 by increasingly smaller numbers. The resulting answers will become larger and larger. Ask students, "Do you think there is a limit to how large the numbers can get?"

n	$20 \div n$
20	
10	
2	
1	
0.5	
0.1	
0.01	
0.001	

• LESSON 10 • Multi-Step Problems

• Objectives

Students will be able to

- use the standard order of operations to find values of expressions with more than one operation;
- recognize how their own calculator processes a series of operations and be able to work with it;
- recognize the equality between the two forms showing the distributive property;
- write single equations describing the multiple steps needed to solve different problem situations.

• Background

Understanding and Recognizing Multi-Step Problems

The concepts stressed in this lesson (order of operations and the **distributive property**) are essential to the early study of algebra. More importantly in this course, they are critical to discussions of multi-step problems. They make up the "ground rules" that govern how multi-step problems are to be written and solved.

Many of your students will have the experience and savvy to know how to use more than one step to solve real-life situations. For example, to figure how much change they should get from a $20 bill, they know to add the prices of their purchases and then to subtract that total from $20. They also are likely to have experience finding averages.

However, to do well on the set-up problems on the GED Test, students must be able to recognize a single equation that combines these two steps. They must be able to see which equation says the same thing as their method. For students who have a good sense of mathematics but are weak in the academic language of math, **this could be one of the single most important lessons in their preparation for the test**.

• Lesson Recommendations

Page **76**

Mental Math Exercises

These mental math exercises ask students to apply what they learned about multiplication and division. In multiplication, as the factors get larger, the answer increases. However, in division, as the divisor increases, the answer decreases. In Exercise 1, since $45 < 50$, the answer to the second equation must be less than 3,750. In Exercise 2, since $95 < 100$, the answer to the second equation must be greater than 57. Ask your students to verbalize the reasoning they used to determine their answer. This will help reinforce the concepts.

Pages **76–77**

Order of Operations

Your class can have some fun as you work together on **Problems 1** and **2**. After pairing students, encourage them to justify their answer choices. When different calculators come up with different answers for the seemingly simple example of $4 + 5 \times 6$, students can appreciate the need for a standard approach to the order of operations.

To be sure that students see the differences among calculators, try to bring a calculator of each kind to class. A scientific calculator will have the order of operations built in, while a simpler one will not. Although most of your students will be using a simple calculator, some will have scientific ones, and students need to be aware of how both will process these problems.

The examples on the bottom of page 76 and the top of page 77 are structured so that you can show how dramatic an effect the incorrect order of operations can have on a problem. Whenever you write something incorrect on the board or overhead projector, be sure to cross it out with a large *X* before you go on to the next problem. Some students listen only selectively and need the extra emphasis to clarify what was correct or incorrect.

Note: The rules presented in this section are only part of the whole set of rules for the order of operations. The **P**lease **E**xcuse **M**y **D**ear **A**unt **S**ally rule will be introduced on page 90 after the introduction of exponents. In the meantime, the student text is very careful *not* to say to do the multiplications and divisions *first*. It says only to do them before the additions and subtractions. However, note that the operation within the parentheses will always be completed first, even after exponents are discussed.

The interpretation of the fraction line as a grouping symbol is usually not discussed until well into an algebra class, but your students will need it in order to understand the formula for finding averages on the formulas page of the GED Test.

Pages **78–79**

The Distributive Property

This is one of the few algebraic concepts represented on the GED Test that is not readily understood just by knowing how to use variables.

Take the time to explain to students how the areas of the rectangles on the top of page 78 represent both the *factored form* and the *expanded form*. After this, your students will agree that the two statements produce the same answer. The next step is for them to be able to convert from one form to the other. Again, the reason they need to know how to do this is to prepare for the set-up problems on the GED Test.

To demonstrate this, present students with the following example:

> Write an expression that shows the total distance traveled by a trucker who averages 55 mph and travels 8 hours one day and 10 hours the next.

Some students may answer with $(55 \times 8) + (55 \times 10)$. While this is correct, they should be prepared to recognize that the answer choice of $55(8 + 10)$ also is correct.

To summarize, the distributive property gives the choice of two ways to do the problem (it doesn't matter how it's done), but students must be able to recognize the equality of both forms.

Page **80**

Using Your Calculator

Some students will do **Example A** as follows:

24 X 11 − 1008 = −744

They need to recognize that the correct answer is +744. Remind them that $a - b = -(b - a)$. This is an acceptable method with these understandings.

Most people are not aware of the different ways that calculators process the same problem. Tell students that the best way to understand what their calculator is doing is to watch the display as they enter a series of numbers and operation signs. Help the individual student arrive at a satisfactory way to use his or her calculator for multi-step problems.

Pages **81-82**

Solving Real Problems with Many Steps

This section should be the primary focus of the lesson. The examples given here are typical of situations that are used as the context of problems on the GED Test. When solving the perimeter and average problems, students can begin with the formulas that appear on the formulas page (page 246). In both cases, however, the student book helps to make sense of the formulas so that problem solving involves critical thinking, rather than just substituting numbers in a formula.

EXTENSION ACTIVITY: **DIVISION AND THE DISTRIBUTIVE PROPERTY**

Finding a mean can be an example of how the distributive property works for division as well as multiplication. When you use an example of finding the average of two numbers, it is interesting to show that you can either add first and then divide by two, or divide each number by two and then add the results.

Example: Find the average of 56 and 42.

Solution A

1. Add first. $56 + 42 = 98$

2. Divide by 2. $98 \div 2 = 49$

Solution B

1. Divide first. $56 \div 2 = 28$
$42 \div 2 = 21$

2. Add the results. $28 + 21 = 49$

Pages **82–84**

Miscellaneous Many-Step Monsters

The remaining problems do not follow any formulas, so begin again with translating from an English expression to a mathematical expression. Before you go on to discuss the examples that are solved in the book, take time to review *when* to add (when combining unlike amounts), multiply (combining equal quantities), subtract (separating or comparing), or divide (separating into equal groups).

As you read each problem, ask the students to interpret the problem situation as a combination of operations. Allow time for them to figure things out. Do not expect answers to be immediate. The middle step (how to plan it) shown in the student book emphasizes that, in complicated cases, one must do a great deal of planning before writing an equation.

Before you do **Problems 10–15** together, have the class discuss general approaches. Then each student should individually try to write an equation for each problem. Ask students to share these equations for class discussion. Let the class discuss and judge whether the different equations are equivalent to the one given as the answer.

With this exercise, you are building your students' skills in recognizing the correct answer in a set-up problem on the test. You are also teaching them to respect methods of solution other than their own. Of course, if an equation is not right, it must be corrected. Keep in mind that you need to maintain an atmosphere where mistakes can be corrected without embarrassment. Even an incorrect response usually contains some element of critical thinking that can be praised.

Page **248**

Another Way to Multiply: The Front-End Method

This page offers students an alternate way to compute the exact answer of a multiplication problem. You can introduce this method during **Lesson 10** as an example of the distributive property. This method, called the front-end method, relies heavily on students' ability to handle trailing zeros (see page 55). The features that follow are what I like about this method.

1. The first partial product is a good estimate of the answer.
2. It helps students picture the process. For many adult students, being able to see a process is necessary for learning. Understanding the process of multiplication is much easier with this method.
3. It has fewer rules than the traditional multiplication algorithm. For example, when multiplying two 2-digit numbers, the order in which you multiply is not important as long as you multiply each digit of one number by each digit of the other number.

   However, this method is not as efficient as the traditional algorithm; it takes up more lines. I personally feel this is a small sacrifice to pay for a method that emphasizes understanding of the process involved. Another reference to this method occurs in **Lesson 13** when we review the estimation technique for multiplying.

LESSON 11 • Powers and Roots

Objectives

Students will be able to

- use the notation of exponents and evaluate expressions containing exponents;
- use the $\sqrt{}$ sign and recognize finding the square root as the inverse operation of squaring;
- follow the standard order of operations as it applies to exponents;
- estimate the square root of numbers that are not perfect squares;
- apply the Pythagorean theorem to find the length of a side of a right triangle;
- apply these skills to find the areas of squares and circles.

Background

The topic of exponents has traditionally been placed at the back of adult education math textbooks, so it is not usually covered in a typical student's preparation for the GED Test. In fact, field-test results show that many examinees resort to pure guessing on the items involving exponents. This is unfortunate because it is very easy to solve the problems, once the notation is understood. The mystery surrounding the exponent and the radical sign have made this topic seem like a foreign language to students. The main thrust of this lesson is to make the notation understandable, to make the student comfortable with squares and square roots, and then to provide uses for both operations by using them in some everyday formulas.

Lesson Recommendations

Page 88

Mental Math Exercises

The topics of squaring numbers and finding square roots are introduced through the operations of multiplication and division.

Page 88

Exponents

The operation of raising a number to a power is introduced as a shorter way to multiply identical factors, just as multiplication was introduced as a shorter way to add identical addends. Students are not required to evaluate the complex expressions, merely to understand what they mean.

For students, the most common misunderstanding involves confusing problems like 3^2 with $3 \cdot 2$. To demonstrate that these are entirely different, show students how to solve each, and then repeat the meaning of an exponent.

EXTENSION ACTIVITY: **COMPARE/CONTRAST**

You can help students to differentiate multiplying factors and raising a number to a power. Give them a comparison/contrast activity. Show them problems like the following example, and have them explain their answers.

Choose the correct answer for each:

		Choices	
$4^3 =$ __	$4 \times 3 =$ __	12	64

Page **89**

Squares

The first and most obvious application of exponents is in finding the area of a square. Tell students that it is no accident that the operation and the shape have the same name. An added challenge is provided by the second example, which introduces converting from one square unit to another. Errors on this kind of problem are often found in the field tests conducted by the GED Testing Service.

Page **90**

Square Roots

Use a variety of approaches to explore the relationship between squaring and finding the square root. To break through some layers of the mystery surrounding the idea of finding square roots, use the example along with the question "What number multiplied by itself is 64?" In my classes, I use the word *unsquare* parenthetically for a time until the student really accepts the words *find the square root.*

To give some extra help to those who still have trouble with the concept, expand on the idea of the inverse operation and equivalent equation. Write $3^2 = 9$ and expect students to respond with $\sqrt{9} = 3$; write $\sqrt{100} = 10$ and expect them to reply $10^2 = 100$; etc.

Page **90**

Order of Operations

With the introduction of exponents and radicals, you now have the full spectrum of operations and can formalize the rules for order of operations in the standard format. We have merely inserted the exponents between the parentheses and multiplication and division. Take time with **Problem 5**. It is designed to help you find any misunderstandings that still remain.

Pages **91-92**

Pythagorean Theorem

It is safe to predict that the GED Test will contain a problem that refers to the Pythagorean relationship between the lengths of the sides of a right triangle. However, it is also likely that the question will involve a Pythagorean triple—whole numbers that fit the relationship. Therefore, the discussion in the student book begins with a definition of the theorem, then talks about these special groups of numbers that "work" nicely in the formula. There is no attempt to prove the theorem; that is beyond the scope of what the GED Test requires. However, the sketch of the 3-4-5 triangle does show it is true for that specific triangle.

Note: Your students may be interested to know that even before Pythagoras proved the theorem, builders in ancient times used its principle. A standard "tool" they carried was a length of rope, knotted at intervals corresponding to the lengths of 3, 4, and 5. They would use this to test whether the beams they were erecting were perpendicular.

The fourth example on page 94 points out one of the pitfalls of teaching the Pythagorean triples. Whenever students see a triangle with two sides being 4 and 5, they think they know that the other side is 3. This is a false conclusion; the missing side in this case is the longest side.

Pages **93-94**

Estimating and Calculating Square Roots

The table of squares given in the student book is only a partial table showing some of the squares that are helpful in estimating square roots. To aid students, remind them of the following points.

1. They know the squares of 1-12 from the multiplication facts.

2. They can figure the squares of numbers such as 20 by using what they've learned about trailing 0s. For example, $20 \times 20 = 400$.

3. Encourage them to memorize special cases such as 13^2, 14^2, 15^2, and 25^2.

The text accompanying the table explains its use in estimating square roots of numbers that are not perfect squares. This process of finding a value between those given in a table is called **interpolation** and is an important skill.

Using a calculator to find the square root of a number is the accepted method in today's world. But to use that tool wisely, one must know approximately what number to expect so that errors can be detected. Calculators are not yet allowed on the GED Test, but students may encounter an item on the test whose answer is not a perfect square. If so, they will see answer choices such as these:

(1) between 7 and 8

(2) between 8 and 9

(3) between 9 and 10, etc.

EXTENSION ACTIVITY: **CONSECUTIVE SQUARES**

You want your students to recognize perfect squares automatically. Therefore, repeated exposure in a variety of settings is a good idea. One possible activity shows the relationship between consecutive perfect squares.

Adding a consecutive odd number to each square results in the next perfect square.

Number	Square
1	1 +3
2	4 +5
3	9 +7
4	16 +9
5	25 +11
6	36 +13

Continue this to the number 15. Divide students into two groups. Then have one group of students continue the process with 16-19 and the other group find the squares by multiplying.

Note: Explain that the difference between consecutive squares is always odd. If one square is even, the next will be odd. See if your students know why.

EXTENSION ACTIVITY: **PATTERNS WITH SQUARES**

This activity shows students the interesting patterns formed by the last digits of perfect squares.

What digits can be the last digit of a perfect square?

$0^2 = \boxed{0}$ 0
$1^2 = \boxed{1}$ 1
$2^2 = \boxed{4}$ 4
$3^2 = \boxed{9}$ 9
$4^2 = 1\boxed{6}$ 6
$5^2 = 2\boxed{5}$ 5
$6^2 = 3\boxed{6}$ 6
$7^2 = 4\boxed{9}$ 9
$8^2 = 6\boxed{4}$ 4
$9^2 = 8\boxed{1}$ 1
$10^2 = 10\boxed{0}$ 0

Notice that the last digits in the answers are **palindromic**. (They read the same backward and forward.) With the students' help, find and record the perfect squares for the rest of the numbers through 20.

Does the pattern continue? Yes. Do any of the perfect squares having two or more digits have all odd digits? No. Do any have all even digits? Yes, $20 \times 20 = 400$.

• LESSON 12 • Circles

• Objectives

Students will be able to

- use the special vocabulary of circles;
- use various approximations for the value of π;
- find the circumference and area of a circle when given either the radius or the diameter;
- discriminate between problem situations that involve area and those that involve circumference.

• Background

Understanding Circles

Experience has shown that the GED Test problems involving circles are among those most likely to be missed by examinees. Devoting this entire lesson to circles is intended to prevent your students from contributing to these statistics.

If at all possible, make this a hands-on lesson. Such an approach will lessen the mystery surrounding circles. Once students actually experience the fact that a circle's circumference is a little more than 3 times its diameter, they will not forget it, nor will they be stymied by the value of π.

• Lesson Recommendations

Page **96**

Mental Math Exercises

These exercises provide a quick review of the meaning of the radical sign.

Page **96**

Radius and Diameter

What is a circle? Precisely speaking, a circle consists only of the curved line and not the enclosed area. The actual definition of the circumference is "the length of a circle." The same principle applies to a rectangle and its perimeter. Realistic models of these shapes would be made of bent wire or plastic drinking straws rather than pieces of cardboard.

Page **97**

Circumference and Pi (π)

Students often memorize the approximate value for π without understanding its origin. The following activity should help students understand the relationship between the circumference and diameter of a circle and π.

EXTENSION ACTIVITY: **INTRODUCING π**

After the class understands the concepts of radius and diameter, provide each group of students with various circular objects of *different* sizes. The purpose of this activity is to compare the diameter of each of these circles to the distance around it. These objects can be trash can lids, margarine tub lids, tin cans, or even a circular cracker.

Since you don't want to get bogged down in actual measurements, provide pieces of string or heavy thread that are cut to the length of the diameters of the circles.

1. Instruct the groups of students to check first that the string actually is the length of the diameter of the circle they have. Make one mark on the edge of the circular region to serve as the starting point for future measuring.
2. Stretch the diameter string out on a desk, and align the starting mark with one end of the string (Fig. 1). It is easiest to measure the circumference by rolling the circle on the desk (or do this on the floor for the larger circles). Roll the circle until you get to the end of the string.

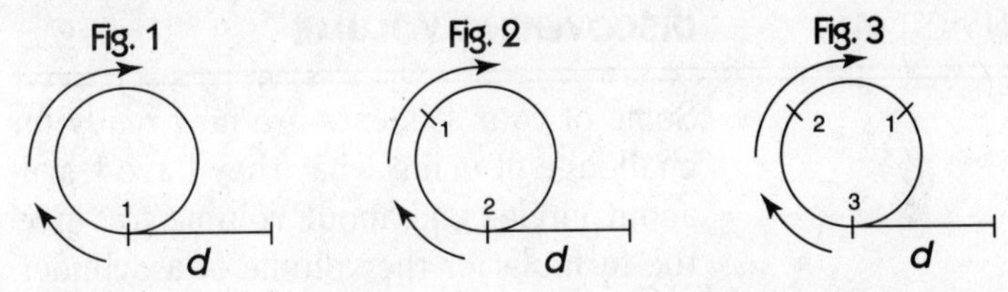

3. Repeatedly reposition the beginning of the string at the place where the string previously ended (Fig. 2). Continue to roll the circle until you get to the end of the string each time. Do this until you come full circle to the starting mark (Fig. 3). By keeping count of the number of times they must reposition the string, the students will find out how many diameters it takes to get all the way around the circle. (After measuring the circumference by rolling the circle on a flat surface, they will also be ready for **Problem 5** on page 97.)

After all of the groups have finished measuring the circumferences, discuss their findings. No matter what the size of the circle was, going around the circle took a little more than three diameters. This is remarkable. In fact, there were ancient cultures that worshiped the circle because of this "mysterious" property.

This activity serves as an introduction to π. Be sure not to refer to 3.14 as a *precise* value for π. All of our computations of the circumference of circles are *approximations*; those using 3.14 are merely closer than those done with 3, and those using the calculator key $\boxed{\pi}$ are even closer than those done with 3.14.

Page **98**

Using Equivalent Equations

The previous activity showed the rearrangement of the formula $C = \pi d$ into $\frac{C}{d} = \pi$. A more useful arrangement of the formula in the real world is $d = \frac{C}{\pi}$. By using this form of the formula, you can find diameters that cannot be measured.

Pages **99-100**

Area

The series of sketches on page 99 is certainly not a *proof* of the formula. (In the fourth picture, the circumference halves are pulled out to be straight lines. This would only be possible if the wedges were infinitely small.) This portrayal does, however, give some visual reference for what is likely a misunderstood concept. As you follow the sequence of pictures, focus attention on the fact that the radius is multiplied by itself. This characteristic of multiplying one dimension by another is always common to finding area (two-dimensional). This point enables students to distinguish between finding area and finding the circumference of a circle.

EXTENSION ACTIVITY: **DISCOVERING VOLUME**

Some of your students are now ready for the challenge of using what they have learned about circles and about volume to "invent" the formula for the volume of a cylinder. Place a cylindrical container next to a rectangular box. Ask the students to remember what they did to find the volume of the box. They found the area of the bottom (number of cubes in the first layer), then multiplied by the height (number of layers). Follow the same sequence with the cylinder. The area of the bottom (a circle) is πr^2, and the height is h.

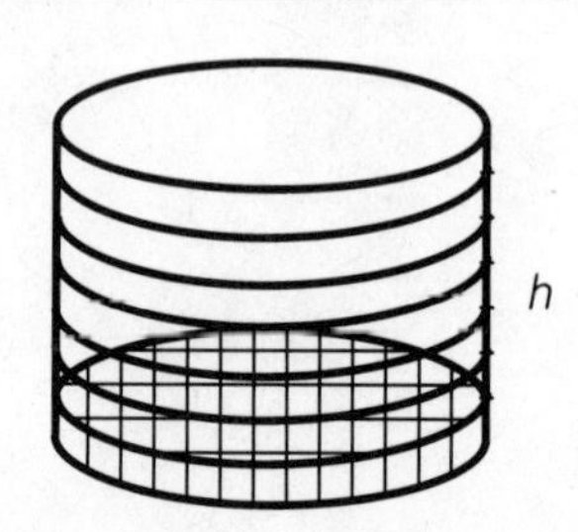

Check with the formulas page to assure everyone that the formula is indeed $V = \pi r^2 h$.

Page **101**

Check Your Understanding

You can expand **Problem 4** into a class project. Have students bring to class the prices of various sizes of pizzas sold at several local pizzerias. (They probably never realized that 12″, 14″, etc., referred to the diameter of the pizza.) Have them figure which size pizza gives the best value per square inch (area). They can also see which size gives the most circumference for the dollar.

LESSON 13 • More Powers—Powers of 10

Objectives

By understanding the decimal system, students will be able to

- write numbers in expanded form;
- multiply and divide by powers of 10 by moving the decimal point;
- write numbers, large and small, using scientific notation;
- solidify their estimation techniques using multiplication and division.

Background

Putting It Together

The broad purpose of this lesson is to build number sense. As students study the basics of place value through the manipulations of scientific notation and as they review and standardize their estimation techniques, they will tie together the isolated concepts they have been learning. It is a timely summary before the checkpoint in **Lesson 14.**

Lesson Recommendations

Page **102**

Mental Math Exercises

The ease of multiplying by powers of 10 is a result of having a number system based on 10. In this lesson, adding the trailing zeros will be correlated to moving the decimal point.

Pages **102–103**

Powers of 10 and Place Value

Students have intuitively used the idea of expanded numbers in previous lessons. Here it is stated explicitly: the value of a digit depends on the place it occupies.

EXTENSION ACTIVITY: **PLACE VALUE AND EXPONENTS**

Handout 4
Place-Value Chart
TG Page 104

Project a transparency of the full-page place-value chart (TG page 104). Place digits in the slots, and ask students to read the resulting numbers aloud.

Example: Place a 7 in the thousands slot and a 5 in the tens slot. Read this number (seven thousand, fifty). Now move these same two digits, putting the 7 in the millions place and the 5 in the ten thousands place. Again read the resulting number (seven million, fifty thousand). Continue this until your class is comfortable with the place names.

In the student book, the three facts on page 103 connecting the number of zeros and digits to the powers of 10 are just observations. Do not interpret them as rules to remember. In further lessons, students will be asked to make these observations for themselves, to notice patterns that occur, and to make some generalizations. This is an example of the mathematical way of thinking that is essential in our technological society.

Page **104**

Multiplying and Dividing a Number by a Power of 10

Again use the full-page place-value transparency (TG page 104) to illustrate the examples. Write the number 49 on an *overlay* transparency, positioning the digits so that the 4 is on top of the 10s slot and the 9 is on top of the 1s slot. When 49 is multiplied by 100 using the mental technique of tacking on the trailing zeros, the result is 4,900. Show this by merely *sliding* your overlay over until the 4 is in the 1,000s slot. Use the same technique for the other three examples as well. It will dramatize how easy these problems are to solve and will provide a better sense of what moving the decimal point is all about.

Page **105**

Scientific Notation

The topic of scientific notation is important for general numerical literacy. We as a nation are accused of not realizing the difference between millions, billions, and trillions as we hear these astronomical figures describing the cost of programs and the budget deficit. The following situation might start students thinking about the relative sizes of large numbers.

If you spent $1,000 a day, it would take about 3 years to spend $1 million. At that same rate ($1,000 a day), it would take you about 3,000 years to spend $1 billion.

The students' simple calculators fail with these large numbers. (For example, enter 800,000 × 1,000,000 and press [=]. The display will show an *E*, the symbol for "error.") However, scientific calculators will display values larger than the screen can hold by converting them to scientific notation. (Enter the same problem. The display will read [8.   11], meaning 8×10^{11}.)

Most of this section will be self-explanatory and a continuation of what students have learned earlier in the lesson. One confusing element is that of negative exponents. Students should know that they exist and can be interpreted as, "Multiplying by a number with a negative exponent is the same as dividing by that same number with a positive exponent."
Refer to **Example B**:

$4 \div 1{,}000 = 0.004$ is the same as $4 \times \frac{1}{1{,}000} = 0.004$

4×10^{-3} $\qquad 4 \times \frac{1}{10^3}$

Pages **106-107**

Estimating with Multiplication

This last section outlines and summarizes the estimation techniques for multiplication and division. It is essential to stress again that there is no one right way to estimate, even for a specific problem. The choice of technique is a personal preference. If students start to compete for closer and closer answers, soon they are taking more time than if they had used their old paper-and-pencil algorithms. Stress that one of the reasons to estimate is to save time.

Introduce the notation of $360{,}000^{+}$ and $380{,}000^{-}$ as a way to close in on the actual answer. Confusion concerning the notation can be avoided by consistent phrasing: the *actual answer* is more (or less) than the *estimate*. Your students will be pleased that they have no trouble estimating the 6 answers in **Problem 11** on page 107 and **Problem 17** on page 109 in less than two minutes.

Pages **108-109**

Estimating with Division

By starting with the denominator and then rounding the numerator to be compatible, the emphasis will be placed on the reason for rounding. The answer to the boxed part of each problem featuring the compatible pair is the first digit of the estimate. The number of remaining digits in the numerator determines the number of remaining digits in the answer.

Many of your students will be happy to learn an alternate way to perform long division. The algorithm that is demonstrated on page 249 is commonly taught in many foreign countries. In trials in the United States, it has been found to be more "learner-friendly" than the method we learned traditionally. Skill in this technique increases as one's estimation powers increase. For this reason, it seems to be a natural method to use in this course, which emphasizes estimation.

The student who is comfortable with the traditional algorithm for long division may notice that the last example corresponds closely to what he or she has mastered as a technique. This solution represents the highest estimation skills at work. It also points out the beauty of this method of instruction: one does not have to attain this level of efficiency in order to get the right answer.

LESSON 14 • Checkpoint II

Objectives

Students will

- review the methods of this unit;
- be aware of some of the formats used on the GED Test;
- evaluate their progress.

Lesson Recommendations

To simulate the actual time allotted on the test, allow only 30 minutes for students to complete the test questions. Use the time before the test to review and redo problems that were troublesome for your class throughout this unit.

Page **112**

Test-Taking Tips

Go over the sample problems on page 113 involving Art's Quik Lube. This will prepare students emotionally for the problem-solving process of applying the basics that they know to new situations that may be unfamiliar. They will also be reminded of the insufficient-information problems, as well as the ones with extraneous information.

When the students are finished with Checkpoint II, discuss the problems and possible solution strategies as you did for Checkpoint I. Encourage students to reveal their personal methods of choosing the answer. Point out that by first taking time to analyze the problem, they could do all the necessary computation mentally and by estimating. Even if they had calculators available, they wouldn't have chosen to use them.

In the time remaining, group the students so they can again play the basic-facts games as described in Checkpoint I **(Lesson 6)**. At this point, the students are responsible for knowing addition, subtraction, multiplication, and division facts with quick recall.

EXTENSION ACTIVITY: **"I HAVE, WHO HAS?" REVIEW**

I learned this game, called "I Have, Who Has?" at an adult educator's workshop in Oklahoma. The game can be modified to review any math skills with any number of people participating. The sample shows 10 participants reviewing the skills we have studied.

Materials: A set of 10 index cards, each containing one of these statements:

I have 1.	Who has 100 times as many?
I have 100.	Who has half as much?
I have 50.	Who has this, divided by 5, plus 2?
I have 12.	Who has twice as much?
I have 24.	Who has this divided by 4?
I have 6.	Who has this times 8?
I have 48.	Who has one more than this?
I have 49.	Who has the square root of this?
I have 7.	Who has this times 9?
I have 63.	Who has this divided by 63?

As you construct sets of cards to fit your class and what you are doing, keep in mind these restrictions: First, no number can be repeated. Second, the series of questions should be circular, ending at the same number where it started. Ideally, there will be one card per person.

Procedure: Pass out the cards, one to each person. Start with anyone. Have that student read his or her card aloud clearly. The person whose number satisfies the question responds by reading his or her card. The next one responds in turn until the process ends back at the starting person. Everyone must remain alert and continue to compute mentally to be able to recognize his or her number when it is described.

Modifications: You should have several sets of cards on hand. The subject matter can fit what you have just studied, or it can provide a review. After seeing the pattern, some students may want to construct their own set for the class.

If the topic is purely review, you can make this game more challenging by giving each student two or more cards.

Pages **114–116**

Check Your Skills

Problem 1: This item asks the examinee to estimate a square root. To answer correctly, students must understand the notation and know the squares of single-digit numbers. Correct answer: **(3)** 8 and 9.

Problem 2: This is an application of the relationship $d = rt$. Students will divide the distance (375) by the time (2.5). Estimating that $\frac{375}{2.5}$ is between $\frac{375}{2}$ and $\frac{375}{3}$ shows that the answer is between 200 and 100. From this, students can isolate the correct choice. They could also have merely noted that the answer had to be more than 100 and that 937.5 was not reasonable. Correct answer: **(4)** 150.

Problem 3: This problem asks how many 400s there are in the area shown. Area can be computed in many ways. Using subtraction, the large rectangular region ($55 \times 50 = 2{,}750$) minus the cutout ($15 \times 40 = 600$) results in 2,150 square feet. There are five 400s in this, but the remainder would require an additional can of paint. Most of this computation can be done mentally. Correct answer: **(4)** 6.

Problem 4: The information "1,050 miles away" is not necessary to solve the problem. To answer the question, students must recognize that they are being asked to *separate* 48 into groups of 7. In this real situation, it is not important that the numbers don't divide evenly. Correct answer: **(2)** $\frac{48}{7}$.

Problem 5: This multi-step problem requires that the total volume (in cubic feet) be multiplied by $.10. The choices correspond to reasoning errors that may have been made in computing the volume. To find volume, students need to multiply 3 dimensions—in this case, $5 \times 5 \times 8$. Mentally they arrive at 200 for this interim value, then must divide by 10 (or multiply by 0.10) to find the correct number of dollars. Correct answer: **(1)** $20.

Problem 6: There are two key parts to identifying the correct answer choice here. First, students must recognize that the fencing around this figure is represented by circumference and therefore requires the formula $C = \pi d$. Second, they must choose the line that is the diameter of the circle. Correct answer: **(3)** 3.14×10.

Problem 7: By recognizing that angle *BOD* and angle *AOB* are supplementary, one just has to subtract 115° from 180° to find the correct answer. The fact that angle *COD* is a right angle is not necessary for the solution. Correct answer: **(3)** 65°.

Problem 8: This multi-step problem would be classified as an algebra problem on the GED Test, but it can be solved using the problem-solving techniques of this chapter. Multiplying $3 \times \$1.25$ gives the cost of the rolls ($3.75). Subtracting that from $4.95 leaves $1.20 for the 2 coffees. A single equation to describe what was done is $c = \frac{4.95 - 3(1.25)}{2}$. Correct answer: **(2)** $.60.

Problem 9: The examinee must recognize that $2 must be multiplied by the number of hours (n), then added to $15. Correct answer: **(4)** $15 + 2n$.

Problem 10: This problem is a challenge for those who do not check their answers for reasonableness.

(1) This is the result of $\frac{92 + 64}{5}$.

(2) and **(4)** contain numbers used in the problem.

(3) This is the average between the two given temperatures.

(5) Correct. Since the problem asks for a five-day average, 5 numbers were needed.

Problem 11: This problem reapplies the work done in **Lesson 5**. The lines meet 1 unit to the right of the origin and 3 units below (1, −3). Correct answer: **(5)** (1, −3).

Problem 12: The single equation that could have been written for this problem is $y = (1{,}950 - 360) \times 12$.

An estimate could be $y \approx 1{,}600(10)^+$ or $\$16{,}000^+$. How much more? Approximately $2 \times 1{,}600 = 3{,}200$.

(1) This is the monthly take-home pay. The examinee needed to go one step further.

(2) This answer resulted from adding instead of subtracting the deductions.

(3) Here the monthly take-home pay was multiplied by 10 instead of 12.

(4) Correct answer.

(5) This answer is unreasonable in this situation. It is more than $12 \times 2{,}000$.

Problem 13: This typical Pythagorean problem asks for the hypotenuse when the legs are 9 and 12. Even if the student did not remember the triples, he or she could have reasoned this way: the missing side must be longer than 9 and 12, but less than $9 + 12$ (the length of the two if joined in a straight line). That leaves two choices, 15 and 18. He or she knows that $9^2 = 81$ and that $12^2 = 144$. Adding these, $144 + 81 = 225$. Notice the last digit of this square number ends in 5. The square root of 225 also must end in a 5. $15^2 = 225$ is also a handy fact for students to know. Correct answer: **(2)** 15.

Problem 14: An understanding of exponents is required in this symbolic problem. If students selected answer choice **(3)**, they multiplied 3 by 2 instead of squaring it. Correct answer: **(4)** 12.

Problem 15: According to the distributive property, there are two correct options to look for among the answer choices: $100\,(6 + 3 + 7)$ or $(100 \times 6) + (100 \times 3) + (100 \times 7)$. Only the first option is listed as an answer choice. Correct answer: **(1)** $100(6 + 3 + 7)$.

Problem 16: By substituting the given values into the formula for the area of a triangle, students can see the puzzle involved in this problem: $A = \frac{1}{2}\,bh$, $30 = \frac{1}{2} \times 5 \times h$. Without doing any algebraic manipulations, they can see that 30 is half of 5 times something. Recognizing that 30 is half of 60, they can determine that since 5×12 is 60, the missing dimension is 12. Correct answer: **(3)** 12 cm.

LESSON 15 • Size of Fractions

Objectives

Students will develop number sense about fractions by being able to

- understand the meaning of a fraction, not only as a part of a whole, but also simply as a division problem;
- recognize which fractions have values close to 1 and $\frac{1}{2}$;
- rename as mixed numbers fractions that have numerators larger than their denominators;
- interchange common fractions with their decimal equivalents;
- recognize equal fractions intuitively when working with rulers, and analytically, using a mathematical principle.

Background

A Different Approach to Fractions

Neither the emphasis nor the methods of this group of lessons on fractions follow the traditional textbooks. There are two reasons for this. First, fractions are becoming less important in problems that occur in everyday life. Now that more calculating is being done by machines, more numbers are expressed in decimal form rather than in fraction form. Second, the old way did not work very well. Students soon forgot which of the manipulation rules to use when. They became more adept at avoiding fractions than at understanding them.

Your students will be taught to carry out the operations with the common fractions that occur frequently in measurement situations. With fractions that are less common and more difficult to picture, they will be taught to estimate.

The main content of this lesson is the basis for understanding the concept of a fraction and its size. In later lessons, this will become the foundation for the estimation procedure and an intuitive understanding of the operations with fractions.

Lesson Recommendations

Page **118**

Mental Math Exercises

The purpose of these mental math exercises is to help students discover how the size of the denominator affects the answer. Point out to students that as the denominators get smaller, the answers get larger.

Pages **118-120**

What Does a Fraction Mean?

The concept usually stressed is the idea that a fraction is a part of a whole. However, the interpretation of a fraction as a division problem is the one that connects what students already know (division) with the application of fractions. Students should be reassured to find that these dreaded fractions are nothing more than the division that they already know.

Review the fact that the order of numbers in a division problem makes a difference. The "problem" $\frac{4}{5}$ is different from $\frac{5}{4}$. Use the same words you used with division to determine the decimal value of $\frac{4}{5}$. For example, you could think of the problem as 4 divided by 5, the number of 5s in 4, or a smaller number divided by a larger number. All should lead naturally to an answer less than 1. Similarly, for $\frac{5}{4}$, students should expect a value greater than 1.

With this kind of buildup, the principle indicating when a fraction equals 1 is obvious to students. **Problems 1-4** provide the reinforcement.

The term *improper fraction* is *not* used to describe fractions with values greater than 1. The term carries a negative connotation that is not appropriate. In fact, these values are often easier to work with when they are expressed as single fractions. However, for comparing the sizes of these fractions, the mixed number notation is easier to use.

Page **120**

Fractions Close to $\frac{1}{2}$

In the following lessons, when estimation is taught, students will need to know not only which fractions are "close to" $\frac{1}{2}$, but also which are greater or less than $\frac{1}{2}$. Be cautious here; students can become confused if they start by doubling the numerator. Instruct students to *begin with the denominator*, find half of it, and then determine whether the numerator is greater or less than half the denominator. If the numerator is greater, the fraction is greater than $\frac{1}{2}$.

Page **121**

Which Fraction Is Larger?

This page is designed to make students feel more at ease with fractions and ultimately to give them more options when they are estimating with fractions. Emphasize that, while they can always use their calculator to find out precisely which fraction is larger, reasoning it out mentally is often much easier (and more satisfying).

The underlying principle of these exercises is the same one students discovered in **Lesson 9**: the larger the number you are dividing by, the smaller the answer.

Pages **122-123**

Decimal and Fractional Equivalents

This section serves a dual purpose. First, students must have these equivalents memorized for later work. Also, now is the time for students to realize how much they already know. The equivalents listed make sense and can always be reasoned out if students forget them. Be open to ways to reinforce these equivalencies. One possibility is to make a set of "I Have, Who Has" cards specifically for this topic.

Pages **123-124**

Equal Fractions

Students need to be aware that there are many fractional names for the same numbers. This is the idea behind equal fractions. I have used a ruler as a concrete representation of this fact because measuring with rulers is one of the most common sources of fraction problems in the real world today. Students need proficiency in using a ruler as well as an understanding of fractions. You can fill both needs by using rulers or tape measures in your discussions, rather than pie drawings or other manipulatives.

Students will gain more from this experience if each of them has (1) an actual ruler to use as a reference and (2) a personal copy of a ruler that has been blown up to a size large enough to write on (Handout 5, TG page 105). You should also have a transparency of the blown-up rulers. Proceed by labeling the inches, then the half inches, etc. (as the student book shows), allowing the ruler to function as a number line for each of the denominations. Have students follow along on their copies, noting the equivalencies as they occur.

It is important that students have a concrete representation of equal fractions to picture in their mind before they are taught the "Fundamental Principle of Equal Fractions." (As you may have discerned, I decided to name this principle for easy reference in later lessons.)

EXTENSION ACTIVITY:

MEASURING WITH FRACTIONS

Take some time to teach your students how to use their rulers to measure actual objects. Expect them to name the ruler marking that each object comes to. You could begin this exercise by using a transparent ruler on the overhead projector to measure some strips of paper. Then ask the student groups to measure some common distances like the space between the lines on their paper, the length and width of their books, and the thickness of their desks.

Students could also practice measuring objects for approximate lengths or widths (for example, to the nearest inch or half inch). Then have students discuss different possible applications for exact and approximate measurements.

LESSON 16 • Adding and Subtracting Fractions

Objectives

Students will be able to

- picture adding and subtracting common fractions on a ruler or a number line;
- use the rules to add and subtract common fractions and mixed numbers;
- estimate answers to addition and subtraction problems that involve fractions other than the most common ones;
- decide whether an estimate is sufficient for an everyday problem.

Background

Nearly everyone's experience is testimony to the fact that people quickly forget the rules of working with fractions. For this reason, the student book gives those rules only secondary emphasis. The primary focus of this lesson is on picturing the problems on a ruler and estimating. So that students can readily picture the problems, each should have a copy of the Fraction Table (Handout 6, TG page 106) to use and write on as he or she wishes.

Lesson Recommendations

Page **126**

Mental Math Exercises

The purpose of these mental math exercises is to review comparing fractions. This is a critical skill for students to have as they begin estimating with fractions.

Pages **128–129**

Adding Fractions

Problem 7 is exploration. Have students actually line up lengths of paper along the expanded ruler to see where the sum falls.

Problem 8 uses only the most common fractions when applying the rule for addition. Students should be getting comfortable with renaming and simplifying these by now. For the fractions (such as the thirds, fifths, and tenths) that are not represented on a ruler, have students use the Fraction Table (Handout 6, TG page 106) to visualize their sums.

Using a transparency of the Fraction Table (Handout 6, TG Page 106), demonstrate **Problem 8h**. Cut a strip of paper $\frac{1}{2}$ unit long. Show this length being added to $\frac{1}{5}$.

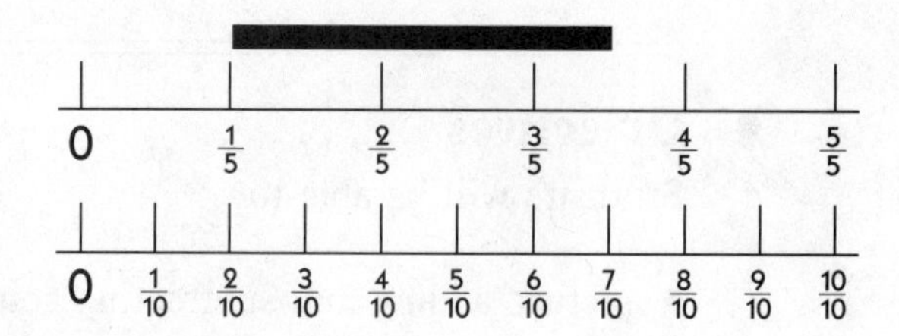

Where does the end of the strip fall? It falls *between* two of the one-fifth marks. To find a fraction that corresponds to this (a tick mark at exactly that point), have students look down on the next number line marked in tenths to see the $\frac{7}{10}$ mark. This dramatizes the need for a least common denominator.

Page **130**

Subtracting Fractions

Begin again with picturing the problems on a ruler. Then use the rule $\frac{a}{c} - \frac{b}{c} = \frac{a-b}{c}$, but do not extend its use beyond the most common fractions. By comparing the rules to the pictures, students can see that finding the common denominator merely assures that all the spaces on the number line are of the same size.

Page **131**

Adding and Subtracting Mixed Numbers

Students must see that a plus (+) sign is understood to be between the whole number and fractional parts of a mixed number. Then they can apply the dependable principles they learned earlier about order and grouping when adding.

Borrowing in subtraction is not stressed here so that the students will focus on visualizing the fractions. In most cases with common fractions, students can picture their way through the problem **(Problem 14c)**.

You can also introduce students to some commonsense ways to think of borrowing. Start with an example like **Problem 14h**, $9 - 2\frac{1}{8}$. Both the 2 and the $\frac{1}{8}$ must be subtracted from 9. Subtract the 2 first, leaving 7. Then subtract the $\frac{1}{8}$ from 7, leaving $6\frac{7}{8}$.

Another way to think of "borrowing" is to *add* the same number to both fractions. For example, try $5\frac{1}{8} - 2\frac{1}{4}$. Renaming leaves $5\frac{1}{8} - 2\frac{2}{8}$. *Choose* to add $\frac{6}{8}$ to both numbers. This makes the second number a whole number. The result, $5\frac{7}{8} - 3 = 2\frac{7}{8}$, is an easy problem to solve.

This is a "compensation" technique often used by mental math whizzes. It also works with whole numbers. This technique is shown in "Another Way to Subtract: Subtracting Without Borrowing," Appendix page 247.

Page **132**

Estimating When Adding and Subtracting Fractions

This section is where students apply the concepts from **Lesson 15**. Not only will students be able to find an estimate that is "close to" the exact answer, but in many cases, they will also be able to tell whether the exact answer will be higher or lower than their estimate.

Again, students should be able to use their common sense along with their developing problem-solving skills. They should be able to come up with answers that are acceptable for most situations in everyday life as well as discriminating enough to indicate the right answer choices on the GED Test.

Underlying the techniques outlined in the student book is the principle that, for every difficult problem, there is an easy one nearby to compare it to. The answer needed should be "close to" (and close enough to) the answer of the easy problem. Encourage your students to be inventive in finding these comparisons. Remind them that the examples in the answer key represent only one possible means of solution.

Page **134**

Adding and Subtracting Fractions in Real Situations

By now, your students are aware of how close to the exact answer they can get by estimating. This should encourage them to estimate more in their everyday lives. This page points out the sufficiency of an estimate in some common situations that involve fractions. Ask your students to recall times in their lives when estimating with fractions would have been adequate. Discuss how this awareness can give a person more control over the numbers in his or her life.

LESSON 17 • Multiplying and Dividing Fractions

Objectives

Students will be able to

- recognize which operation to apply to fractional situations ("of" means to multiply);
- multiply and divide fractions, canceling when possible;
- estimate when finding fractional parts of a number by rounding to a compatible number.

Background

Confusion arises with multiplication and division of fractions mainly because students believe that when they multiply, the answer must be larger than the original number. Although they have already seen multiplication by a number less than 1 **(Lesson 9, page 72)**, they will need to be reminded that when multiplying by a number less than 1, they should expect a smaller answer.

Lesson Recommendations

Page **138**

Mental Math Exercises

The purpose of these mental math exercises is to reinforce students' understanding of the relative size of fractions.

Page **138**

Finding a Fraction of a Fraction

The sketches *show* students that when they multiply by a fraction less than 1, they should expect an answer that is less than the original number. This result also seems more likely in problems when the word *of* is stressed (for example, $\frac{1}{4}$ *of* $\frac{3}{8}$ is less than $\frac{3}{8}$). Reemphasize this point with each problem in **Problem 1**.

Pages **139–140**

Finding a Fraction of a Number

Finding fractional parts of numbers is the most common everyday use for fractions. For example, students already know the fractional parts of an hour. Additionally, it is closely related to finding a percent of a number, the topic of later lessons. Expect your students to master these skills.

If your students can cancel, they can make these problems even easier. However, be sure to point out that failing to notice the opportunity to cancel and multiplying the original numbers does not make the answer wrong. They will just have to simplify the answer.

Pages **140-141**

Estimating Fractional Parts

Canceling is very important when estimation and mental math are stressed, as they are in this book. Canceling can make the difference between students needing paper and pencil and being able to find the answer in their head. Students will like canceling. In fact, some students like it so much that they try to use it all the time—even when adding and subtracting.

Remind your students regularly that canceling is a technique allowed only when multiplying. Of course, this includes being able to cancel in a division problem after finding the reciprocal of the divisor. By this time, they have changed the division problem into a multiplication problem.

While following the delicatessen example on page 140, have students compare each estimated answer with the precise calculator answer. The next lesson includes instruction on how to do some fractional problems on a calculator, but using a calculator now can show that estimation is often close enough.

Pages **141-142**

Multiplying Mixed Numbers

Some of your students will be able to recognize the distributive property of multiplication over addition in these sketches and examples. The fact that the sample problem $3\frac{1}{5} \times 1\frac{1}{2}$ is expanded as $(3 + \frac{1}{5}) \times (1 + \frac{1}{2})$, using *both* addition and multiplication signs, means that one cannot just group the whole numbers and fractions as for addition of mixed numbers. Instead, the addition in the parentheses must be done before the multiplication.

Note: With small numbers, the addition step is accomplished by making the mixed number a single fraction. The multiplication step follows, then the answer is changed back to a mixed number. However, this method will result in tedious computations when the numbers get large. For this reason, as well as the fact that the fractional parts lose their relative importance when the whole numbers are large, I recommend using estimation for these problems.

Page **143**

Dividing Fractions

Students have already learned that when they divide by a number less than 1, the answer should be greater than the original number **(Lesson 9, page 73)**. Asking the question "How many __s are there in __?" also justifies the size of the answers in division problems. By using these two aids, your students should be able to see the reasonableness of the answers to division problems. Have students note that dividing by 2 is the same as multiplying by $\frac{1}{2}$.

Page **144**

When to Multiply or Divide

This page expands students' ideas about *when* to multiply and divide. They will continue to multiply when they combine equal groups **(Problem 10)** and divide to separate into equal groups **(Problem 13)**. But now they also have to multiply when they are asked to find a fractional part of something. A second interpretation of division is also added (it was mentioned in earlier lessons also): divide when the problem can be interpreted as asking, "How many __s are there in __?" Encourage your students to use this question as a kind of test to decide whether to divide in problems that do not meet the conditions they learned earlier with respect to equal groups.

LESSON 18 • Making Connections

Objectives

By reviewing the relationship between fractions and decimals, students will be able to

- learn the decimal equivalents of the eighths;
- find the answer to problems using either decimals or fractions;
- see the connections between the ways the operations are completed with fractions and decimals;
- explore the relationship between original price, discount, and sale price, and solve problems involving these quantities;
- recognize the options available when solving multi-step problems involving fractions.

Background

We cannot expect our students to synthesize the parts of a course of study into an organized whole without some help. The purpose of this lesson is to aid students in comparing the new information from this chapter on fractions with what they did in prior lessons with decimals. In some cases, their knowledge of fractions will help them understand why the decimal methods worked, and in other cases, the opposite will be true. When students understand the comparisons, they will have more options for solving problems. They will become more flexible problem solvers. So, while this lesson could be considered a review, it also has an important purpose of its own.

Lesson Recommendations

Page **146**

Mental Math Exercises

The purpose of these exercises is to help students discover patterns and connections in their math knowledge.

Pages **146-147**

Decimals and Fractions

Although memorizing the decimal equivalents of the eighths is not as important for students as memorizing the thirds, fourths, and fifths, there are occasions when this knowledge would be helpful. Develop the values for these equivalents by using the Fraction Table transparency (Handout 6, TG page 106).

Starting with the first number line, label each tick mark with its decimal equivalent. There should be no hesitation naming these before you get to the line marked off in eighths. On this line, label the equivalents students know because they are identical to the previous line.

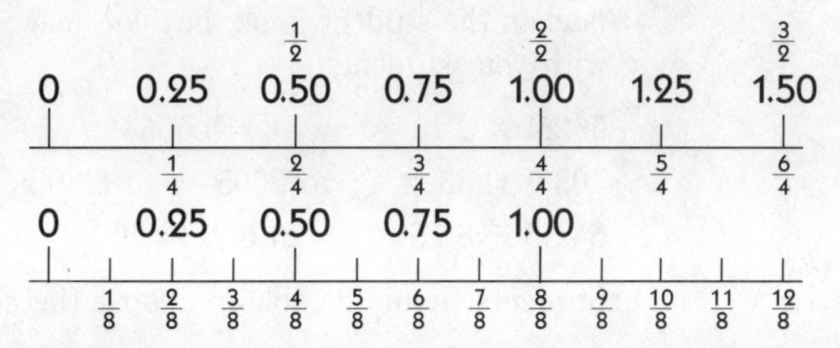

From the Fraction Table, it should be obvious to students that $\frac{1}{8}$ is half of $\frac{1}{4}$ and that to find the decimal equivalent of $\frac{1}{8}$, they need to find half of 0.25. To find the equivalent of $\frac{3}{8}$, students need to find the number halfway between 0.25 and 0.50, etc.

How do you check these results? Enter the fraction into the calculator as a division problem: $\frac{1}{8} = 0.125$, $\frac{3}{8} = 0.375$, etc.

The example problems show that, by having options, students can choose the method of solution that is easiest for them at that time. The ready availability of a calculator may be the reason one student chooses to use it to find the answer. But a student who first has to find the calculator may choose to do the problem without it. Your students are beginning to have enough experience with calculators that they can be objective about using one. A calculator is not the cure-all they had hoped it would be. Students can see some disadvantages to using one.

When fractions are involved in a problem, there are new considerations to weigh in deciding how to find the answer. Except for multiplication with fractions, entering a fraction problem into a calculator correctly is more work than doing it with paper and pencil. Encourage your students to try to use their calculator, learning about memory keys and parentheses, to solve addition, subtraction, and division problems. Then tell them to use their judgment in choosing the most appropriate method.

Note: At the time of this writing, calculators are not being allowed during the GED Test. However, the problems on the test are not the type that would send a person scurrying to find a calculator. Tedious computations are not required on the test. The problems and observations in this lesson are meant to convince your students that they can do this kind of problem *without* a calculator.

Problem 2 introduces another commonly used estimate for π. It is best used when the diameter is divisible by 7, so that canceling is possible.

Pages **148-149**

Fraction and Decimal Operations

Comparing the methods for addition, subtraction, multiplication, and division of fractions with the methods previously used (but perhaps not specifically taught) for decimals will validate both sets of procedures. This helps students understand why they did what they did and also helps them remember what to do when. If they forget, there's a good chance they can figure out the procedure from their understanding of the other form.

The student book has not formalized the rules for multiplying and dividing by decimals until now. You can mention these rules in a way that makes sense to students. To place the decimal point in the answer of multiplication and division problems, students can also use estimation. This will be mentioned later in the student book, but you may wish to try some of these problems with your students:

$562.4 \times 2.11 =$ a) 118.6664 b) 1186.664 c) 11866.64

$103 \times 0.35 =$ a) 3605 b) 360.5 c) 36.05

$54.35 \div 8.75 =$ a) 6.2114285 b) 62.114285 c) 6211.4285

Students should be able to choose the correct answer by estimating.

Pages **150-151**

Multi-Step Problems

Your students can discover the relationship of original price = discount + sale price if you help them complete a table such as the one on the right:

$\frac{1}{4}$ **OFF EVERYTHING**		
Original Price	Discount	Sale Price
$ 20.00		
28.00		
40.00		
80.00		
100.00		

Have students use mental math to fill in the blanks. After the table is complete, ask students to look across each row of the table, looking for a relationship. Ask them to verify that the relationship they have discovered works for every row. Finally, ask them to express this relationship as a formula. Be prepared to accept as correct all four of the equivalent statements: $o = s + d$, $o = d + s$, $o - d = s$, and $o - s = d$, where o equals original price, d equals discount, and s equals sale price.

Some of your students will be able to see that the two methods for solving these problems are really just examples of the distributive property. Either multiply first $[240(1 - \frac{1}{4}) = 240 - (\frac{1}{4} \times 240)]$ or subtract first $[240(1 - \frac{1}{4}) = 240(\frac{3}{4})]$. In the discussion of percents, this idea will be developed in more detail.

LESSON 19 • Checkpoint III

Objectives

Students will

- review the methods and concepts of this unit;
- evaluate their progress.

Lesson Recommendations

Page 154

Test-Taking Tips: Multi-Step Problems

A common error that examinees make while taking the GED Test is that they choose a response that is the same as the answer they get while doing the first step of a multi-step problem. The example cautions against this and then shows that the partial (or interim) answers will likely be among the choices given. Students who are aware of this ploy are less likely to be tricked by it.

To simulate the actual time given for the GED Test, allow students 35-40 minutes to complete the Checkpoint questions. Please stress that students should use mental computation and estimation whenever possible. Also note that this set of items is unusually high in the percentage of items that involve fractions. On the official published practice tests, only about $\frac{1}{10}$ of the items actually require knowledge of fractions.

Use the class time remaining to discuss the problems and some possible ways to find the answers. Again, most problems require only mental computation or estimation, while none are of the type that require tedious computations.

EXTENSION ACTIVITY: **"I HAVE, WHO HAS?" REVIEW**

If you have more class time, play a round of "I Have, Who Has?" using the new concepts. (Refer to the explanation given in Lesson 14 on TG page 55.) For example, this set of 10 index cards would be appropriate:

1. I have 12, who has $\frac{1}{2}$ of this?

2. I have 6, who has this squared?

3. I have 36, who has $\frac{1}{4}$ of this?

4. I have 9, who has 1 more?

5. I have 10, who has $\frac{1}{10}$ of this?

6. I have 1, who has $\frac{1}{4}$ of this?

7. I have 0.25, who has 3 times this?

8. I have $\frac{3}{4}$, who has 4 times this?

9. I have 3, who has this times 6?

10. I have 18, who has $\frac{2}{3}$ of this?

You may wish to write the problems on the board as they occur; sometimes it is difficult to picture canceling without the written problem.

Pages **155-159**

Check Your Skills

Problem 1: This item is a multi-step problem where the interim answer of \$3.45 is response **(1)**. The correct choice, **(4)** \$6.55, is the result of subtracting the interim answer from \$10.

Problem 2: By knowing the decimal equivalent of $\frac{1}{8}$, students are able to choose the correct answer, **(2)** $\frac{1}{8}$ in.

Problem 3: The word *yards* is underlined for emphasis. Students must know that m yards is the same as $3m$ feet. They then multiply length by width to find the area of $9m$ sq. ft. Correct answer: **(4)** $9m$.

Problem 4: This question asks for the answer to $\frac{2}{5}$ of the capacity. But a number for total seating capacity is not given. Correct answer: **(5)** Not enough information is given.

Problem 5: By multiplying the whole number and then the fraction by 5, students would get $5\frac{5}{4}$. In its simplest form, the correct answer choice is therefore **(4)** $6\frac{1}{4}$ yd.

Problem 6: Algebraically, one would add 7 to 17, then divide by 3. However, substituting the given answers also is effective. Correct answer: **(4)** 8.

Problem 7: Students must know that the sum of the angles in a triangle is 180°. If they subtract 30° from 180°, then divide by 2, they find 75°. The problem contains extraneous information, which should be ignored. Correct answer: **(3)** 75°.

Problem 8: A batting average is determined by dividing the number of hits by the number of times at bat. This was explained in **Lesson 18**. A problem like this would not be on the actual test without that information. Correct answer: **(2)** .750.

Problem 9: Students should know the decimal equivalent of $\frac{1}{10}$. Correct answer: **(2)** 5.35 km.

Problem 10: \$889,000,000 is written as $\$8.89 \times 10^8$. Correct answer: **(4)** 8.89×10^8.

Problem 11: Students could have found 12 on the x-axis, then gone straight up until they hit the line of the graph. They should have then moved evenly across to the left until they met the y-axis. They would land between 8 and 10, so the answer is 9.

A second method involves using the formula. Substitute 12 for the number of pounds of cans: $d = 0.75 \times 12$, or $\frac{3}{4} \times 12 = 9$. Correct answer: **(3)** \$9.

Problem 12: This time, students should have started on the y-axis at 10. They should have then moved evenly across to the right until they hit the line of the graph, then straight down to find the number of pounds. They would land between 12 and 14. The only possible response is **(4)** $13\frac{1}{3}$.

A helpful hint for problems like these: Tell students to overlay a corner of their paper on the graph itself to ensure that they move straight across as well as straight up and down. The corner should fall precisely on the line of the graph, and the edges should be perpendicular to the axes of the graph.

Problem 13: By putting the eighths in decimal form, students can determine that $7\frac{7}{8}$ (7.875) is the highest rate. Students could also use estimation. Correct answer: **(3)** $7\frac{7}{8}\%$.

Problem 14: Only one answer makes sense if students ask, "How many $\frac{1}{8}$s are there in $2\frac{1}{2}$?" Correct answer: **(1)** 20.

Problem 15: This problem represents the most difficult of the fraction problems that could make it onto the test. Not only does it require many steps, it requires division by a fraction. If any of your students get it correct, they deserve accolades.

After the 12 acres of roads and parkland are subtracted, 28 acres are left for houses. The problem asks, "How many $\frac{2}{3}$s are there in 28?" If students remember that dividing by a number less than 1 results in a larger number, they will have only 2 choices remaining—**(4)** and **(5)**. Any number of mental strategies (in addition to actually dividing) could be used to choose 42. Correct answer: **(5)** 42.

Problem 16: Angle *BDC* and angle *EDC* are supplementary. Students need not know this term in order to see that the two angles together form a straight angle—180°. Correct answer: **(4)** 135°.

Problem 17: All of your students should have succeeded in choosing the correct answer since the right-angle symbol is used in the diagram. Correct answer: **(2)** right.

Problem 18: Students could automatically eliminate choices **(4)** and **(5)** because the answer must be less than 1. They should then estimate that $\frac{100}{127} \approx \frac{100}{125}$. That would reduce to $\frac{4}{5}$. Correct answer: **(3)** $\frac{4}{5}$.

Problem 19: This problem is another example of the distributive property. The correct choice shows the addends already added ($4\frac{1}{2} + 1\frac{1}{2} + 3 = 9$). If your students were tempted by choices **(1)** or **(5)**, they were depending too heavily on the physical appearance of a problem without being careful of the signs. This is the second time this temptation has been used, and it won't be the last time they see it! Correct answer: **(2)** 9(8.50).

Problem 20: Technically, this problem asks students to multiply $30 \times 1\frac{1}{6}$. However, students with common sense can find the answer more easily. They know that the answer has to be more than 30. How much more? How far would the cyclist travel in 10 minutes? Picture a round clock. All the way around equates to 30 km, halfway around would be 15 km, and a third of that (10 minutes) would be 5 km. Correct answer: **(5)** 35 km.

• LESSON 20 • Comparisons: Fractions as Ratios

• Objectives

Students will be able to

- write ratios describing real situations;
- recognize when ratios apply to situations;
- appreciate the wide spectrum of situations for which ratios are applicable;
- express ratios in simplest form;
- recognize relationships present in a proportion;
- use the Law of Proportionality to determine whether two fractions are equal.

• Background

Connecting Prior Knowledge with Ratio and Proportions

Your students already have a strong background as they begin this chapter on ratio and proportion. They learned the relationship between multiplication and division by using two "rate" relationships, $d = rt$ and $c = nr$. They know that the line in a fraction means "divided by." This chapter on ratio and proportions builds on these concepts; it does not start over with a new topic. Take every opportunity to point out to students those concepts that they have already seen.

The techniques the students learn in this lesson are, without a doubt, the most practical math skills they will ever learn. Many everyday problems can be solved by setting up a proportion and finding the missing value. When the GED Testing Service asked people across the country to send in their ideas for good items to put on the math test, a majority of respondents sent in items that used ratio and proportion. Generally speaking, 40 percent of the items that appear on the test *can be solved* using these techniques. For some students, there are easier ways to find the answer, but the relationship is there. The message is clear. Your students must learn these techniques! They are extremely practical, and they can help your students pass the test.

This lesson serves as an introduction to ratios and proportions. This includes applications, the relationships between numbers in a proportion, and the fundamental Law of Proportionality. In later lessons, this law will be the foundation for solving proportions, including percents. Because everything in this lesson is so practical, your students will find it easy and enjoyable.

Lesson Recommendations

Page **160**

Mental Math Exercises

The purpose of these exercises is to review what students know about equivalent fractions and how to manipulate fractions.

Page **160**

Using Fractions to Compare

Introduce your students to three ways to write a ratio. Of course, the most mathematical way (fraction form) is the one we will pursue. Also note that there are different ways to represent the same relationship. In the example, the student/teacher ratio is $\frac{11}{1}$ while the teacher/student ratio is $\frac{1}{11}$. Both are valid ratios for the relationship. To determine which to use in a problem, students must look for the order of the words in the problem. If they use the same order as the words, they will be correct.

Page **161**

Rates

For students to be able to recognize when ratios are appropriate, they need to analyze how they are used. Rates do not necessarily involve time, but they always involve the idea of "for each" or "per." In **Problem 2**, the students are asked to find this relationship in various situations. Note: Some authors make a distinction by referring to ratios as comparisons between like units and rates as comparisons between unlike units.

Page **162**

Ratios Compare by Division

Not all comparisons can be expressed with a ratio. You can make this apparent to your students by making some observations about the room you are in. If there are 8 students and 12 student desks, you could compare the number of students to desks by using the ratio $\frac{8}{12}$. On the other hand, you could also compare the number of students to desks by saying that there are 4 fewer students than desks. This statement compares by subtraction and does not involve ratios.

EXTENSION ACTIVITY: **REAL-LIFE RATIOS**

Here is an example of ratios found in the news:

Recent findings link the likelihood of heart attacks to the amount of "waist fat" a person carries. They say that a man whose waist-to-hip ratio is greater than 1 has a higher risk of having a heart attack than the general population. For women, if the waist-to-hip ratio is greater than 0.8, she is in the danger zone for heart attacks.

Although it could prove embarrassing to do measurements in the classroom, this medical tidbit is a great example of how a ratio can be the most effective way to compare two quantities. Because of the varying sizes of individuals, researchers couldn't say something like "if a woman's hip size is less than 10 inches more than her waist size," because small women would panic. It is the size of the ratio that is critical, not the measurements themselves.

Page **163**

Percent

The topic of percent is introduced as a rate here. Students need to feel comfortable with the easy parts of this concept before moving on. However, since percent is such a commonly used rate, your students may surprise you with the amount they understand.

Problems 4d and **e** are meant to be explorations of how much students can reason using percents. If your students do not have difficulty with these, go further than the student book and ask how many *i*'s they would expect in 100 letters, 200 letters, and 1,000 letters.

Page **163**

Slopes

Slopes of lines are another of the topics in geometry that are made to seem difficult by a complicated formula ($\frac{y_2 - y_1}{x_2 - x_1}$). This introduction avoids the coordinate plane by talking about the slope triangles that can be sketched.

Page **164**

Expressing Ratios in Simplest Form

In ratios as well as in probability, simplifying fractions is an important skill. From this point on, your students should understand that when a ratio is required, it should always be in simplest form.

EXTENSION ACTIVITY: **EVERYDAY RATIOS**

Ask your students to look in their newspapers or magazines for examples of ratios being used to compare numbers. They will find countless examples of percentages, but ask them to look further for other simple ratios. These can be anything from 4 out of 5 doctors surveyed to gale-force winds of 80 mph.

Pages **164–165**

Equal Ratios Are Proportions

The student book points out three relationships that exist between the four numbers in a proportion. They are noted so that students will have options when they solve problems. The first relationship is merely a review of how to build equal fractions by multiplying the numerator and denominator by the same number. You can refer to this as a "between" relationship because it exists *between* the two fractions.

The second relationship exists between the top and bottom of the same fraction. That relationship is the same in the other fraction of the proportion. You can call it the "within" relationship because it exists *within* the fraction. This is the relationship one uses in solving analogies. It is not always useful as a method for solving mathematical proportions, but it can be used as a mental check to see whether the answer seems right.

The third relationship is important enough to be a principle. The Law of Proportionality will be used to solve proportions as well as percent problems. It states that if two ratios are equal, their cross products are equal. In this lesson, it is introduced and used to determine whether two fractions (ratios) are equal.

LESSON 21 • Proportions

Objectives

Students will be able to

- set up a proportion that reflects the relationships in a real situation;
- find the missing number in a proportion;
- use proportions to solve problems involving similar figures;
- use ratios to determine the better buy by comparing unit prices.

Background

Practical Ratio and Proportion Techniques

This lesson continues to expand students' awareness of the practicality of ratio and proportion techniques. The method of cross multiplication is used to solve proportions. However, students (and the teacher) must remain "loose" about this technique. From the start of a problem, they must be aware that there are many correct ways to set up a proportion. (The answer key shows only one possibility.) After the proportion is set up, the students should be encouraged to notice different patterns that may help them to solve the problem in an easier way than the structured algorithm of cross multiplication.

Lesson Recommendations

Page **168**

Mental Math Exercises

Your students now know another way to determine the equality of fractions. Encourage them to use more than one method on these problems.

Page **168**

Proportions in Real Problems

In the example, the proportion expresses the cost in cents rather than dollars. This way of setting up the proportion eliminates the decimal points. If a student questions why we can do this, you will have the opportunity to turn the question around and ask the class for the reasons. The Fundamental Principle of Equal Fractions allows us to multiply both the numerator and denominator by 100.

Problems 1h and **i** not only teach students to be alert to the different questions that can be asked, but also introduce the concept of deducing a missing value. If 2 out of 5 people are men, the students must deduce that 3 out of 5 are women. If the ratio of men to women is 2 to 3, they must deduce that the number representing the total is 5.

Page **169**

Finding a Missing Number in a Proportion

After a review of solving equations involving multiplication, this section introduces the algorithm of solving proportions by cross multiplication. While students are encouraged to use their calculator whenever they wish, you should always point out ways to do problems without a calculator. Method 2 at the bottom of the page shows how canceling makes the numbers easy enough to work with mentally.

Page **170**

Fractions and Mixed Numbers

Some of your students (the ones with lingering "fraction fear") will want to change the fractions to decimals in these proportions. Of course, that is acceptable. However, those who use decimals may need to use a calculator, while those who use fractions can solve these problems mentally.

To guard against the "mindless mimicry mathematics" that has dominated math instruction and learning, students are encouraged to step back after writing the proportion and examine it for other possible ways to find the answer. If the relationships are not obvious to them immediately, instruct them to look for the possibility of simplifying the ratio. If no apparent pattern emerges, then they can resort to cross multiplying. It is the "old, dependable" method that they can always use.

Page **171**

Using Proportions to Solve Real Problems

The examples on this page show students that they can apply the method of proportions to problems they have solved previously. Setting up the proportion correctly and carrying out the cross multiplication ensures choosing correctly between multiplication and division. For some of your students, this method will give them a safe, structured way to solve rate problems.

Page **172**

Similar Triangles

Congruency is mentioned only as a comparison for similarity. The word *congruent* remains on the list of geometry vocabulary recommended by the GED Testing Service, but I think your students will need only an awareness of the word.

When figures are similar, their angles are equal, but their sides are proportional. In the second example, the fact that the shadows are measured at the same time and that both the tree and the post are assumed to be perpendicular to the ground are what ensures that the angles in the two triangles are indeed equal.

Page **174**

Best Buys: Comparing Prices

There are so many more things to say about comparing prices than what is taught on this page. For example, your students could set up the ratios differently, comparing the differences in number of ounces to the differences in price.

You could also discuss that, even though the larger container usually has a smaller unit price, there are some items for which you would still buy the smaller container. Why buy a large box of raisins if you end up throwing half of them away because they have hardened?

EXTENSION ACTIVITY: **REAL-LIFE PRICES AND COMPARISONS**

All of the examples on page 174 came from one fact-finding trip to the supermarket. If your class is small and students seem interested in this topic, organize a shopping trip to compare prices. If this is not feasible, compare the food advertisements that appear from the different markets in your city. You can also make students aware of the advantages of using coupons and shopping on double-coupon day, as well as the folly of impulse buying.

EXTENSION ACTIVITY: **HANDS-ON COMPARISONS**

Some of the problems in the ratio and proportion lessons referred to the color of certain candies. You can give your students hands-on experience with ratios and proportions by using packages of multi-colored candies. (This activity also provides a rare opportunity to eat the instructional materials when the activity is completed.)

You can have students compare one package to another. Is the total number of candies the same in each package?

You can also have students compare by colors. What is the ratio of red candies to the total? Which color of candies is the least represented? What is the ratio of red candies to the candies of the least represented color? Is the ratio consistent between packages?

Compare packages of different sizes. What is the total number of candies? Compare the unit price of the candies in each size package. Use the ratios found in the small package to predict the number of a certain color of candies in the large package. Do the color ratios remain consistent?

Keep a record of your findings, and use the data to figure percentages after you have completed the next lesson.

Note: This activity is mentioned in the Massachusetts "Priority Math" materials and was also recommended by a group of teachers at a workshop in Iowa. I have also read accounts of it being used in grade schools where hands-on experiences are being emphasized over rote learning.

LESSON 22 • Percent I

Objectives

Students will be able to

- solve, on an intuitive basis, easy percent problems either in their heads or with a few pencil strokes;
- solve more complicated percent problems by using either proportions or equation solving;
- write a percent equation that describes a problem situation.

Background

Relating Percents and Decimals

The only new concept introduced in this lesson is the relationship between percents and decimals. Percent means "hundredths" as well as "out of a hundred." The solution methods used in this lesson are adaptations of procedures that the students have already learned. You should continually emphasize these connections so that students see mathematics as a unified subject rather than a hodgepodge of unrelated rules and "tricks."

Build on the foundation you have worked so hard to lay. Not only will it make the new learning easier, but it will also strengthen the underlying principles when students see more applications for them.

Several recurring threads appear in this lesson:

- The fraction and decimal equivalents;
- "Of" meaning times;
- Solving proportions;
- Solving equations that indicate multiplication.

Lesson Recommendations

Page **176**

Mental Math Exercises

The purpose of these exercises is to prepare students for the next step of connecting fractions, decimals, and percents.

Pages **176-177**

Quick and Easy Everyday Percents

This introduction to percents is an intuitive approach that is intended to leave students with a broad overview of the concept. It is also intended to show students that many of the most often-used percents can be dealt with conceptually and without formal algorithms. Students should be able to handle these problems mentally after the extensive groundwork laid in the fractions lesson. An additional benefit of these pages is that students will be left with certain benchmarks by which they can judge the reasonableness of answers in more complicated problems.

A survey of the official practice tests shows that many of the percent problems that occur involve these "easy" percents and a solid understanding of the concept. Try to establish these *before* students get caught up in rules and procedures.

EXTENSION ACTIVITY: **CREATING ADDITIONAL PROBLEMS**

If your students respond well to the problems on pages 176-177, you can expand this section by challenging them to devise easy ways to find, for example, 26% of something (by adding 25% and 1% of it), or 9% of something (by subtracting 1% from 10% of something).

Be sure to remark on how powerful students' mental skills have become. This way, they won't panic if they're without their calculator or paper and pencil.

Pages **179-182**

Writing the Percent Equation as a Proportion and Solving Percent Problems Using Equation Solving

Two methods or algorithms for solving percent problems are presented. You may choose to teach both methods or only one. If you choose to teach both (my recommendation), proceed straight through the pages in the order given. However, if you think that more than one method will be confusing to your students, choose one of them and concentrate on it.

If you choose the proportion method, start at page 179 and proceed through page 181, but omit page 182. If you choose the equation-solving method, omit the second half of page 179 (after the percent equation is described) and the first part of page 180 (up to "Writing the Percent Equation for a Situation").

Both methods use the *percent equation*. Students must be able to distill the information from a word problem into the template of this equation before they find the answer. As your students read a word problem involving percents, instruct them to focus on the base. Teach them to ask, "percent *of* what?"

The following problem was placed in this guide instead of the student book because it lost its impact by requiring that the answer be provided merely on paper. It is meant to be a discussion question that dramatizes the importance of the *base* in percent problems. You can use it either at the beginning of class or at the end, wherever it will generate the most discussion. It represents the kind of question that asks students to think for themselves rather than to depend on established rules.

Which Would You Choose?

$10 OFF or 10% OFF

If a student's answer to the question is "It depends," he or she already understands an important aspect of working with percent. Percent is a rate that means "out of a hundred," but it does not exist by itself; percent is always "of" something. To answer the question above, consider the base of the percent. Ask students these three questions.

When does 10% equal $10?	Only when the price of an item is exactly $100.
When is 10% less than $10?	When the item costs less than $100.
When is 10% more than $10?	When the price is more than $100.

LESSON 23 • Percent II

Objectives

Students will be able to

- estimate with common percents, including problems involving money;
- find simple interest and determine the rate of interest charged;
- solve percent of increase/decrease problems.

Background

Working with Percents

This lesson, like the first percent lesson, contains some intuitive and conceptual work as well as the more formal work on percent of increase and decrease. Most examples show both methods (using proportions and using equations) of solution. Students need not be able to do both. Instruct them to follow the example they understand best.

Lesson Recommendations

Page **186**

Mental Math Exercises

The purpose of these exercises is to remind students of the link between fractions and percents. Also, students should have different methods of solution available to them when solving percent problems.

Pages **186–187**

Estimating with Easy Percents

This section extends the mental work of the last lesson. Students can estimate only if they know the benchmark percents and fractions established in the last lesson. Finding compatible numbers for division is especially crucial here. Your students may need a brief review of compatible numbers before you begin.

While completing **Problem 1**, ask the students to indicate whether each actual answer will be greater or less than the estimate.

It is especially important to take the time to talk about how valuable these estimation skills are in everyday situations. For example, a person who can estimate is never surprised at the amount that comes up on the cash register. Talk about estimating before going to pay the cashier. Students should be able to estimate discounts as well as the amount of tax that will be added.

Page **188**

Simple Interest

Point out to your students that the way simple interest is computed in this section is really not the way most interest is computed in today's society. With the aid of computers, banks pay **compound interest** (interest on the principal and on the interest earned to date). However, the basic idea of paying for the use of money is introduced here.

Fractional and decimal percents also are introduced in the context in which they most often occur.

Pages **190-191**

Percent of Decrease/Percent of Increase

Many everyday percent applications are essentially two-step increase/decrease problems. On the GED Test, these are classified as algebra problems.

EXTENSION ACTIVITY: **STRESSING THE BASE OF THE PERCENT**

Copy the following problem situation on the board, excluding the answers given in parentheses. Have students think about and discuss the situation in small groups. Then fill in the blanks as a class. You may need to do a quick review of the terms being used.

A jeweler sells some items at twice the price he paid for them.

selling price = ____% of wholesale price (200%)

profit = ____% of wholesale price (100%)

wholesale price = ____% of selling price (50%)

profit = ____% of selling price (50%)

LESSON 24 • Relating Rates and Slopes to Graphs

Objectives

Students will be able to

- find the slope of a line on a grid by graphing methods or by use of the slope formula;
- graph an equation that represents a simple proportion;
- derive an equation when given a table of values;
- connect the ideas of rate and slope of a line;
- review coordinate graphing, negative numbers, and graph reading.

Background

Understanding Slopes

The main reason for including this lesson is to provide a visual reference for the ideas of ratio and proportion. The link between graph lines and rates is the concept of the slope of a line. This lesson starts with the definition students already know: slope $= \frac{\text{rise}}{\text{run}}$. Here they will be asked to determine the number of units in the rise and the run, first by counting the squares on a grid and then by subtracting the values of the coordinates on a graph.

The lesson then introduces some real-life representations of slopes. The **slope** of a line illustrates the *rate* at which one variable changes with respect to the other variable. For example, on a graph of $d = rt$, the slope of the line represents the *rate* of speed (how fast distance changes with respect to time).

Your students will be exposed to three ways of showing the relationship between variables: the equation, the table of values, and the graph. They can often find the answers to the questions by using any one of these three.

Lesson Recommendations

Page **194**

Mental Math Exercises

The purpose of these exercises is to review what students know about integers. Students are then prepared for the applications of integers presented in this lesson.

Pages **194–195**

More About Slopes

The fact that the slope of a straight line remains the same regardless of which two points on the line are chosen is an example of proportional ideas. All the ratios are equal; they can all be simplified to the same ratio.

You can eliminate a lot of confusion by having students focus on the slope of a line from left to right. With this in mind, if a line leans downward, it has a negative slope. If a line leans upward, it has a positive slope. A horizontal line has a slope of 0. And finally, a vertical line has an undefined slope because 0 is the denominator. (Never divide by 0.)

Pages **196–197**

Finding Slopes from Coordinates

Students are now introduced to the slope formula from the formulas page. Even though the subscripts give this formula an aura of mystery, the underlying process that the formula describes is not difficult at all. Show your students that each of the subscripted variables represents one of the coordinates of the two chosen points.

If the formula proves to be too abstract for some students, they can always refer to the visual graphing method of counting the units to find the slope.

The following issues may require further discussion:

- It is not important which point is chosen to be point 1 and which is point 2. However, the order of subtraction must remain *consistent* for both the rise and the run.
- This book has not covered the rules for multiplying and dividing with negative numbers. Those rules are mentioned in the examples, where the sketches of the lines corroborate the results.

Pages **198–199**

Graphs That Are Pictures of Equations

The equations to be graphed in the student book are limited to those that reflect direct proportions. Notice that the values in the tables can be expressed as proportions, thus reapplying what was learned in **Lesson 20**. Because the mechanics remain simple, students can concentrate on the main ideas. The graph, the table, and the equation all represent the same relationship. By seeing the connection, students can solidify their understanding of each.

Be sure to emphasize that the slopes of the lines in these graphs represent the rates involved in the equations. This concept is critical in the study of graphing equations in algebra and beyond.

Students will be reviewing the graph-reading skills that were introduced in **Lesson 9**. They will use them again in **Lesson 27** with informational graphs.

Page **200**

Comparing Graphed Equations

Graphs that present data over time are introduced here by this graph of a race. Emphasize the comparison of slopes of the two lines; the steeper slope represents the greater rate of speed.

EXTENSION ACTIVITY: **GRAPHING A STORY**

Many of your students will be intrigued to discover graphs telling stories. You can whet their appetites even more by showing the following graph, which tells the story of one person's commute to work:

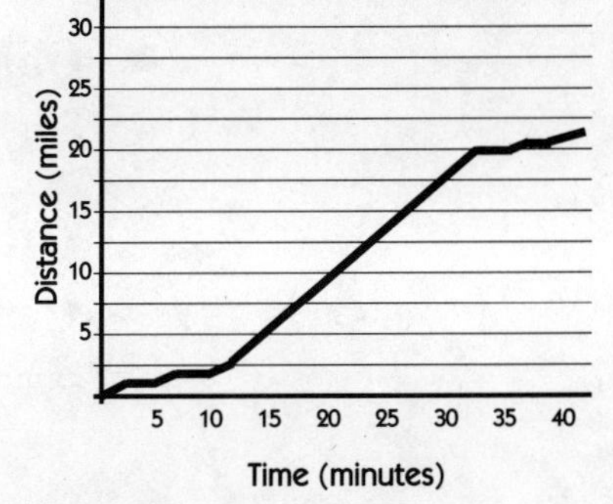

Red lights and stop signs provide occasions for the zero slopes, and highway travel accounts for the faster rates (steeper slopes).

Encourage your students to construct a graph to tell the story of some personal trip of their own or to show the story of a duel between two cars in a race. Pit stops (zero slopes) may be the occasion for one car to pass the other (here the lines would cross).

Another graph of interest, especially at tax time, is this one showing the different *rates* of taxation for the different income ranges. It is accurate enough to use to estimate the amount of taxes you will have to pay if you know your taxable income. (Only the tax credits affect the total tax found here.) Now, why doesn't the IRS include such a graph in the income tax booklet so that people can check their computations? Comparing the schedule to the graph, ask your students to find the slope of the line at each interval.

Tax Rate Schedules

Schedule X—Use if your filing status is Single

If the amount on Form 1040, line 37, is: Over—	But not over—	Enter on Form 1040, line 38	of the amount over—
$0	$19,450	. . . . .15%	$0
19,450	47,050	$2,917.50 + 28%	19,450
47,050	97,620	10,645.50 + 33%	47,050
97,620	. . . . . .	Use Worksheet below to figure your tax.	

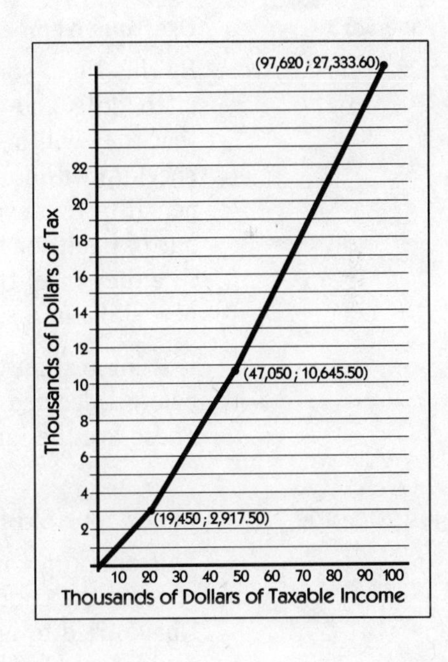

LESSON 25 • Checkpoint IV

Objectives

Students will be able to

- review the concepts of this unit, using the test format;
- evaluate their progress with respect to mental math skills and estimation.

Lesson Recommendations

Pages **202–203**

Where Does the Decimal Point Go?

The sample problems and the text address an uncertainty that is common among learners: "Where do I put the decimal point?" By talking about this problem openly here, you may be able to get your students to share their own experiences with it. Ask them if there are other types of situations in which they are confused about where the decimal point goes. In most cases, the remedy for their confusion will be to use common sense rather than to depend on the memorization of rules.

Page **203**

Use Your Number Sense

By this time, your students should have gained enough practice with mental math skills and estimation that they have some confidence in their abilities. So that they will be forced to use these skills, they should try to complete this checkup without using scratch paper. Many students pick up their paper and pencil to find even the most obvious answers without thinking. This may be out of habit, or they may feel a sense of security from using their old methods. In either case, this checkpoint will give them the opportunity to test their new skills in a situation that is nonthreatening.

It is interesting to note that I constructed the problems in the checkpoint before I decided to make this suggestion. This is not a setup for students. These are the kinds of problems they can actually expect on the test.

Pages **204–209**

Check Your Skills

This checkup contains 28 items, so it is equivalent to half of the GED Test. It stresses the concepts that were covered in this unit and also provides a review. I have tried to balance this set of problems so that it has the proper percentage of algebra and geometry items.

For your convenience, the algebra and geometry items have been classified in the following problem-by-problem analysis. This may give you some insight into the classification of problems. Not all algebra problems necessarily look like the algebra you remember. If manipulation of an equation is involved (such as in a proportion), the problem is algebra.

Problem 1: The information that there are 64 ounces in the package is extraneous. The answer is found by subtracting: $100 - 93 = 7$. Correct answer: **(1)** 7%.

Problem 2: ALGEBRA. No knowledge of the algebraic rules is required. Students can just picture −11 on the number line and move 18 units to the *right* because they are adding. It may help some students organize their thoughts by writing *low* + 18 = *high* and then substituting. Correct answer: **(3)** 7°.

Problem 3: The fact that this is a more complex number with a decimal point does not change how students should approach this problem. The number lies between the perfect squares of 36 and 49. Correct answer: **(2)** 6 and 7.

Problem 4: This two-step problem parallels what many students have already seen. However, this time it is a set-up item requiring them to know the notation for the procedure. Correct answer: **(5)** $20 - 8(1.37)$.

Problem 5: GEOMETRY. The alternative answer choices make it imperative that students remember the order of the coordinates and the role of positive and negative signs on graphs. Correct answer: **(5)** $(3, -2)$.

Problem 6: GEOMETRY. This problem tests the students' understanding of the vocabulary of circles. Correct answer: **(3)** 4.

Problem 7: By estimating that the circumference is about 3 times the diameter, students can choose the correct response. Correct answer: **(2)** 12.56.

Problem 8: The fact that the ride is 20 minutes long is not necessary to find the answer. This problem tests the placement of the decimal point, as discussed earlier in the lesson. Correct answer: **(3)** $11.60.

Problem 9: To solve this problem, students can apply the formula from the formulas page. The underlined words help students to know that they are not finished when they find the amount of annual interest. Correct answer: **(4)** $9,000.

Problem 10: ALGEBRA. This is a two-step problem whose procedures are presented in a single expression. The percentage will be determined by setting up the ratio of $\frac{\text{percent}}{\text{base}}$. Correct answer: **(1)** $\frac{20 - 16}{20}$.

Problem 11: ALGEBRA. To derive the proper equation, students must remember that *of* means times when dealing with fractions, $\frac{1}{16}$ of $w = \frac{1}{16} \times \frac{w}{1} = \frac{w}{16}$. Correct answer: **(2)** $s = \frac{w}{16}$.

Problem 12: GEOMETRY. These two angles combine to make a straight angle, so their sum is 180°. Correct answer: **(2)** 45°.

Problem 13: GEOMETRY. Setting up the proportion for the sides of similar triangles requires three of the four values. Only one is given. Correct answer: **(5)** Not enough information is given.

Problem 14: To solve this problem, students need estimation and a sense of percent. They may have thought, "20% of a hundred is 20, so 20% of 50 is 10." Or they could have found 10% of each item, then doubled each amount. Correct answer: **(3)** c only.

Problem 15: ALGEBRA. Students could use a proportional equation to find the answer. Correct answer: **(4)** $87\frac{1}{2}$%.

Problem 16: GEOMETRY. This is a simple application of solving proportions. Correct answer: **(2)** 32 in.

Problem 17: If students are able to use $\frac{1}{3}$ hour instead of 20 minutes, they can easily figure their answers mentally. Or they may be able to multiply $20 \times 10\frac{1}{2}$ mentally and then divide by 60. Correct answer: **(4)** $3\frac{1}{2}$ hours.

Problem 18: ALGEBRA. Because the numbers are so easy here, this problem can certainly be solved mentally. Correct answer: **(3)** $322.50.

Problem 19: Some students may be confused here, thinking that 9 is approximately $\frac{1}{9}$ of 80. While this is true, the question asks about the electricity used, not the time. These are inversely related. Correct answer: **(3)** 9.

Problem 20: Being able to use $\frac{1}{3}$ hour for 20 minutes is helpful here again. Students could also reason that if a heater uses 1 kWh in 20 minutes, it then uses 3 kWh in 1 hour. 3 kWh $\times$ 8 hr. = 24. Correct answer: **(1)** 24.

Problem 21: This set-up problem requires a careful reading of the problem to notice that there are three traffic lanes in each direction. Correct answer: **(4)** $6(12) + 16.4$.

Problem 22: Common sense should stop students from selecting choice **(5)**. No blueprint would be that large. Correct answer: **(2)** 1:18.

Problem 23: ALGEBRA. Technically speaking, this problem uses the distributive property $(2(n + 3n))$. However, if they are comfortable with the use of a variable, students can add their way around the rectangle. Correct answer: **(5)** $8n$.

Problem 24: Students may use either the table or the graph to choose the correct answer here. Correct answer: **(3)** 10.

Problem 25: ALGEBRA. Remembering how they labeled proportions so that the top and bottom entries correspond should help your students to find the correct proportion here. Correct answer: **(2)** $\frac{4}{12} = \frac{0.80}{c}$.

Problem 26: Again, the graph and the table could both be helpful here. Students should not have to actually set up a proportion for this problem. Correct answer: **(5)** $6.40.

Problem 27: GEOMETRY. This Pythagorean theorem problem uses an easy set of Pythagorean triples. Correct answer: **(3)** 10 ft.

Problem 28: ALGEBRA or number relations. This problem could be solved with proportions (algebra) or just by simplifying the ratio. Correct answer: **(4)** 0.02%.

LESSON 26 • Data Analysis

Objectives

Students will be able to

- collect and organize data using a scatter plot;
- find the mode, median, mean, and range of groups of numerical data;
- read and interpret scales and bar graphs;
- read statistical information with an appropriate amount of skepticism.

Background

Data Analysis Is a Process

This lesson and the next lesson of the student book are a step-by-step exploration of the process of data analysis. By being involved at all stages of the process, your students will gain insights into how important each step can be. By organizing raw data from a survey and then creating a graph illustrating the results, your students will get hands-on experience with the details involved. Rather than having you tell them about it, they do it. An important result of doing it themselves is that the process loses its aura of mystery, and they can readily see how data can be manipulated to tell different stories. Students will know what to look for when statistical information is presented.

One step (planning the survey) was necessarily done in advance. However, if you feel comfortable with the steps of the process, you should feel free to skip the questionnaires in the student book. You and your class could begin a new investigation, asking some questions that are meaningful to the group. Then you could follow the steps in the book, organizing and graphing the relevant information.

If no other topic grabs the attention of your class and you wish to stay with the car questionnaire, you can still skip the answered questionnaires and the "canned" graphs in the book. This would be an advantage especially if the responses given in the book are not typical of your community. To make this approach work, try to get between 20 and 30 responses from people distributed among the age groups. Have students organize and graph the data as was done in the book. The collection of data could serve as a comparison to the numbers in the student book, and your class could try to interpret the differences and speculate on the reasons for them.

Lesson Recommendations

Pages **210-211**

Collecting Information

The construction of the questionnaire is an important factor in interpreting survey results. Questions arise as to whether the respondents were led to choose one answer over another simply by what was previously asked. A second concern arises when the respondents are asked to choose among various answers. What if the answer choices do not cover all possible responses? Both of these weaknesses occur in this questionnaire.

These difficulties will make themselves apparent to students later in the lessons. At this time, you should ask the students whether they notice any problems and how they would change the questionnaire to eliminate these.

Three empty questionnaires are included on page 211 so that each student is encouraged to get involved from the start. The extent to which the additional data is used depends on you. (Refer to the Background section on TG page 89.) At the least, students should see their data in comparison to the book's data.

Pages **212-213**

Organizing Information

This exercise with scatter plots should serve as a review of how to plot points. The completed scatter plot gives an overview of the data and gives the researcher some ideas as to which comparisons might prove interesting to follow up.

Pages **214-217**

Analyzing Numerical Data

These pages show the three established measures of central tendency, or averages. In our usage, the word *average* has come to mean the *mean*. However, each measure is valid in certain situations. **Problems 6** and **7** are important to show how valuable the median is when there are a few outlying points that skew the distribution and distort the mean.

The importance of interviewing a *random* sample is shown in **Problem 13**. By answering the questions, students can become aware of how unscrupulous survey takers can affect the data they collect. This dishonesty has been uncovered in many advertising claims. A second characteristic of the sample—that it should be large enough—was more difficult to translate into an activity. You could make this point by comparing each student's individual data to the means and medians of the entire group.

The fact is that the mean is the most widely used of the averages. **Problems 14-17** are set up so that students can discover some valuable generalizations about this measure. Remind them that they already learned much of this material in **Lesson 9** when they explored the question of "What happens when you divide?"

Pages **218–219**

Presenting and Interpreting Data: Bar Graphs

The problems parallel the examples so that students can mimic the examples. The objective here is that students are finally able to read and interpret a graph. The mechanics of actually constructing one are made as easy as possible so they do not get in the way of that goal.

Problem 26 warns students against making value judgments about the data represented in a graph. Conclusions must be based on the facts alone, not on one's personal view of the reasons for them.

Page **220**

"Numbers Don't Lie"

The easiest way for people to "lie" with statistics is to distort the scale of a graph. The scale is labeled, but often the intent is to deceive those who are not careful readers.

• LESSON 27 • Circle Graphs and Probability

• Objectives

Students will be able to

- collect and organize data using a frequency distribution chart;
- prepare data showing how a whole is divided for entry in a circle graph;
- read and interpret circle graphs;
- use the basic principles of probability in interpreting data from a circle graph;
- read statistical information with the proper amount of skepticism.

• Background

The Importance of Circle Graphs

Information that is appropriate for a circle graph is different from the numerical data graphed in the last lesson. It is important to note that circle graphs show how the total is divided.

By experiencing how the real data from a survey can be distorted, students can gain the skepticism that is so important in being "numerate" when confronted with the facts and figures.

• Lesson Recommendations

Pages **222–223**

Organizing Data

These pages guide students through the steps of organizing and processing data so that it can be presented on a circle graph. As with the bar graph, the mechanics of graph construction are presented but not stressed. The connection with fractions and ratios can serve as a review and validation of students' earlier work. However, the goal is to be able to read and interpret a circle graph, and that should be easy after students see how it is constructed.

One weakness of the questionnaire is exposed in **Problem 2**. The psychological effect of having just been encouraged to name your dream car and then having to come back down to earth and choose a realistic next car would be to highlight the importance of affordability. Remind students that when reading the results of polls, they should look for the actual questionnaire that was used. These will appear with the results of legitimate surveys. Bring examples of both kinds from your newspaper.

Pages **223-225**

Graphing the Data: Circle Graphs

The work that students have done with angle measures will be reinforced here and will show students an application for this knowledge. The two circle graphs that the students are asked to construct in **Problem 10** can be created by "eyeballing," without the use of a protractor.

Pages **226-228**

What Are the Chances?

This three-page discussion of probability centers on aspects of games of chance. The concepts are clear in this context because in games, the outcomes are "equally likely." This is not as true or as easy to see in real-life applications.

Stress that the survey on page 228 was not scientifically designed or administered, so the results are not likely to be valid. However, to show how results like these are used to predict future results, we assume that they do represent the opinion of the population. Decisions are often based on the results of informal surveys such as the one in the student text.

Problem 23 points out a second weakness of the questionnaire. Many people do choose a car because of the status it supposedly brings to them. However, this reason was not one of the choices on the questionnaire, nor was there a place to check whether one's reasons were "other" than those given. The respondents to this questionnaire were forced to choose among the reasons provided. It is obvious that this can distort the data. Remind your students again to carefully read the questionnaires that accompany statistical information before they accept the results as truth.

Page **229**

Probability of Independent and Dependent Events

Independent and dependent events provide the basis for much of the public's misunderstanding of the "laws of probability." Most people believe that it is more likely for a coin to turn up tails after 9 successive tosses that turned up heads. However, this is a misinterpretation. Probability says that, over the long run, you can expect to see an equal number of heads and tails (with a fair coin). But it does not say anything about the *next* toss. For a more detailed and interesting discussion of probability and the public misconceptions of it, see *Innumeracy* by John Allen Paulos, a bestselling book in paperback (Vintage, 1990).

With state lotteries becoming so popular, I felt that some mention of the astronomical odds involved in winning must be made. An informed public can then choose whether or not to participate. The method involved in figuring the odds for the lottery is more complex than the scope of the book, but I will provide it here in case someone asks:

$$P(6 \text{ out of } 49) = \left(\tfrac{1}{49} \times \tfrac{1}{48} \times \tfrac{1}{47} \times \tfrac{1}{46} \times \tfrac{1}{45} \times \tfrac{1}{44}\right) \times \tfrac{720}{1} = \tfrac{1}{13{,}983{,}816}$$

You multiply by 720 because the order in which you pick the numbers doesn't matter and there are 720 (6!) ways to order 6 numbers.

EXTENSION ACTIVITY: **MAKING PREDICTIONS**

You can do this activity as a follow-up to the activity suggested with multi-colored candies in Lesson 21 (TG page 77) or as a new activity stressing statistics and probability.

Let each student or group of students have one bag (all the same size) of multi-colored candy pieces.

- *How many pieces of candy are likely to be in one bag of this size?*

1. Let each group count the number of candies in its bag.
2. Before sharing their information, hold up an unopened bag of the same size. Ask each group to predict how many candies will be in this bag. (The predictions may all be different.)
3. Using the data from each group, find the mean, median, and mode for the number of candies for the entire class (your sample).
4. Again ask how many candies are in the unopened bag.

Which of the three measures of central tendency do your students think is best to answer this question? (Probably the mode.) If the mean is not a whole number (for example, 20.2), it certainly is not the best estimate.

Note that any one of the three averages is a more valid estimate than any one of the group's previous estimates. Increasing the number of bags in the sample makes the estimate more valid.

- *If you were to reach into a full bag and pick out one candy, what is the probability that it would be red? green?*

Repeat the process of the previous question whereby each group finds its own fractional parts without consulting with the others. Then pool everyone's information about the frequency of reds and greens, and find the fractional parts of the entire group of candy. Again ask the question about the full bag. This time the probability estimate will be more accurate.

Have students try an experiment to test their numbers. Have them choose 1 candy at a time from a full bag (without peeking) until they have removed 10 candies from the bag. Is the number of red ones and green ones what everyone would have expected?

LESSON 28 • GED Practice Test

Objectives

Students will be able to

- diagnose their own areas of weakness and complete the necessary review before attempting the actual GED Test;
- develop confidence about their success on the test.

Lesson Recommendations

Pages **231–232**

Test-Taking Tips

The tips on these pages are reminders of some of the most important tips given earlier in the student book. In addition, as a last resort, I provide one of the standard tips for multiple-choice test takers: try each of the alternatives and see which answer works.

At this point, the most effective review for the GED Test is to try the test in the student book. It can serve as a valuable diagnostic tool, review organizer, and confidence builder.

Pages **233–245**

Final Checkpoint

After all the tests have been graded, analyze the results. Find the test problems that only about half your students missed. Let these be the focus of your group review time. This tactic will be the most effective for two reasons:

1. The students who got the problems right can share their strategies with the others. By reminding them that the method of solution in the back of the student book is only one way to solve a problem, you will encourage your students to be inventive in their approach to problems. You want to bring out the nonacademic, nonalgorithmic, commonsense reasons for why they chose the right answers. When students compare their strategies, some will gain more insight, and most will gain confidence.
2. Your intuition may tell you to start your review with the problems that all the students missed because these are areas of real trouble. However, these may be the kinds of problems for which no amount of preparation will help. Now is not the time to confuse students with the more obscure topics.

   In group review, concentrate on the basics. Your students need to know that there will be "foreign" problems on the real test also. Encourage them to use their time wisely by being sure they have correctly answered the problems they are familiar with. There will certainly be enough of those so that they can pass the test. On an individual basis, encourage the students to ask for explanations of other problems that they still don't understand.

The only remaining obstacle to clear is the big one of test anxiety. Everyone experiences it to some degree, and there are many personal remedies. Some students say that a cup of coffee will allow them to think faster. Others say that they get "wired" by coffee and can't think systematically. Personally, I find that nothing is as effective as the relaxed attitude that confidence brings. This is why the least difficult items are at the beginning of the GED Test. Success breeds success. Encourage your students to skip problems that seem difficult to them and return to those problems later after they loosen up their thinking processes with easier problems.

The 56-item test in this lesson has been structured to be as close to an actual GED Test as possible. However, it is not guaranteed to accurately predict who can now pass the actual GED Test. For this purpose, you should use the Official Practice Tests published under the auspices of the GED Testing Service. The following standard score equivalents are reprinted just as a reminder of the number of correct responses (out of 50 graded items) needed to achieve the passing score on the math test in various states.

Warning: Do not interpret these scores precisely for the test in the student book. They are intended only as general guidelines.

Required standard score	Required number of correct answers
45	30
40	22
35	16
30	11

Post-GED

I am confident that many of your students will be turned on to math as a result of the student text's commonsense approach. Once they have passed the GED, they will want to continue and conquer that next hurdle—algebra. Technically, they are ready for a basic algebra course as taught in many community colleges. They will be able to handle variables and can certainly master the sensible word problems. (However, as you recall, in the traditional algebra course, *sensible* is not the adjective you would choose to describe word problems.)

Investigate course offerings at your local community college so that you can recommend certain courses and instructors. Your students are still fragile learners; many cannot withstand the pace and rigor of a one-semester algebra course without some help. I would not recommend a self-paced, individualized course either. Most of your students will need a two-semester course taught by someone who can explain things in everyday English. Help them to find it.

Item Analysis

Problem 1: NUMBER RELATIONSHIPS. Emphasize that memorization is not necessary if students understand the meaning of a fraction. Students can reason that 62% is a little more than $\frac{1}{2}$. Correct answer: **(4)** $\frac{5}{8}$.

Problem 2: ALGEBRA. The relative value of negative numbers is introduced in traditional elementary algebra courses. Students must be able to figure out how much each tick mark represents. Correct answer: **(3)** -1.

Problem 3: MEASUREMENT. This is a two-step problem requiring multiplication first and then addition. This would be a good opportunity to review the order of operations. Correct answer: **(4)** 52.

Problem 4: MEASUREMENT. Another two-step problem. Students need to multiply before they subtract from $20. Stress the ease of using estimation in this problem. Correct answer: **(4)** $6.92.

Problem 5: ALGEBRA. This problem requires the manipulation of the formula for the area of a square. Correct answer: **(2)** between 10 and 11 feet.

Problem 6: MEASUREMENT. Students can choose the order of the multipliers that is easiest for them. Here the ability to handle the trailing zeros is important. Correct answer: **(4)** 6,000.

Problem 7: GEOMETRY. This is a classic proportion situation involving similar figures. Correct answer: **(3)** 80 cm.

Problem 8: MEASUREMENT. The base of the percent is missing. Warn your students that this is a likely scenario for insufficient information on test items. Correct answer: **(5)** Not enough information is given.

Problem 9: ALGEBRA. The use of parentheses in establishing the correct order of operations is being tested here. Correct answer: **(4)** $500 - (169 + 283)$.

Problem 10: DATA ANALYSIS. This problem is a standard type requiring students to find the mean. It is interesting to note that while an easy way to estimate the answer is not apparent, the numbers are very easy to manipulate. Correct answer: **(2)** 240.

Problem 11: ALGEBRA. This is the familiar proportion problem. Students need to be careful where they place the decimal point. Correct answer: **(4)** $3.25.

Problem 12: DATA ANALYSIS. Recognizing that the entire circle of a circle graph represents 100% provides the insight needed to find the answer. Correct answer: **(3)** 40%.

Problem 13: MEASUREMENT. To get the correct answer, students must choose the correct percentage (5%) from the graph, then solve the simple problem 5% of $2,500 = x$. Correct answer: **(3)** $125.

Problem 14: ALGEBRA. Explore the variety of ways your students solved this one. As in the answer key, they could have used the proportion method, or they could have noticed that $5 \times 50 = 250$, then multiplied 3 by 50 to find x. Correct answer: **(5)** 150.

Problem 15: NUMBER RELATIONS. This set-up problem requires students to be able to write a subtraction problem based on time order. Correct answer: **(1)** $d - 85 = 308$.

Problem 16: NUMBER RELATIONS. This problem tests the students' number sense with percents. If they have developed this sense, they would not consider actually finding 25% of each value given. It might be interesting to find out how many of your students did just that. Correct answer: **(5)** b and c only.

Problem 17: MEASUREMENT. Reading tables and graphs is a critical skill. Most students should have no trouble with this problem. Correct answer: **(3)** Duluth.

Problem 18: NUMBER RELATIONS. Find out how many students used the table and how many used the graph to solve this problem. Once students observe both means of solution, which seems easiest to them? Correct answer: **(5)** Detroit.

Problem 19: DATA ANALYSIS. This problem will provide students with an opportunity to use their knowledge of integers and their ability to read tables. Correct answer: **(1)** 18°.

Problem 20: MEASUREMENT. This multi-step problem asks students to extract information from a diagram. Correct answer: **(2)** 3.

Problem 21: MEASUREMENT. This problem involves the students' knowledge of a calendar as used in measuring time. If they *don't* understand that fully, they will have chosen **(5)** as the answer. The time-saving aspects of estimation can also be demonstrated here. Correct answer: **(2)** 232.8.

Problem 22: MEASUREMENT. Whether to multiply the 7 by the 2 or the 3 or neither is the question that will cause the most difficulties for many students. Correct answer: **(5)** $3(12) + 2(24) + 7$.

Problem 23: GEOMETRY. Students could either write an equation using the information in the problem or use the graph to solve the problem. Correct answer: **(3)** 135.

Problem 24: MEASUREMENT. Again, students could write and solve an equation or use the graph to solve this problem. Estimation would be very appropriate. Correct answer: **(1)** 22.

Problem 25: NUMBER RELATIONS. This problem may catch some students off guard. Students just need to understand what the problem is asking for and then use their common sense. Correct answer: **(2)** the total number of hours you parked in the lot.

Problem 26: ALGEBRA. This proportion problem would be a snap, except for the fractions involved. Ask your students how they doubled $1\frac{3}{4}$. Did they break it up into 1 and $\frac{3}{4}$, doubling each of them, or did they go even further and break it up into 1, $\frac{1}{2}$, and $\frac{1}{4}$ to double these? Did anyone actually solve the problem algorithmically, $\frac{7}{4} \times \frac{2}{1} = \frac{14}{4} = \frac{7}{2} = 3\frac{1}{2}$? Correct answer: **(5)** $3\frac{1}{2}$ cups.

Problem 27: MEASUREMENT. Most students will be sure that they have to multiply 3 by 64 to get the number of cans. After that, knowing that the number of cases is fewer than the number of cans should lead them to the correct choice. Correct answer: **(3)** $\frac{64 \times 3}{24}$.

Problem 28: GEOMETRY. After sifting through all the extraneous information, students should realize that the only important fact is that the two angles are supplementary, adding up to 180°. Correct answer: **(3)** 135°.

Problem 29: MEASUREMENT. A number of difficulties could arise here. First, students could use the diameter instead of the radius in the formula. Or they could multiply by 2 instead of squaring. Finally, the change in the order of the factors could confuse them. Correct answer: **(2)** $10^2\pi$.

Problem 30: GEOMETRY. The vocabulary of circles is important here. Correct answer: **(3)** *AD.*

Problem 31: NUMBER RELATIONS. Being able to compare the size of decimals is necessary in this problem. Correct answer: **(1)** 3.02 in.

Problem 32: DATA ANALYSIS. This is an elementary problem in combinatorics, a branch of mathematics that was not covered in the student book. I used this problem to show students that they should expect a few "foreign" problems on the test, even though they have completed this book diligently. Help students realize that they do, however, have the skills needed to solve this problem. Correct answer: **(3)** 15.

Problem 33: MEASUREMENT. Students must first recognize that the 5 is extraneous. Then the ability to handle the trailing zeros will lead to success on this problem. Correct answer: **(2)** 9,000.

Problem 34: DATA ANALYSIS. This problem gives a twist to the typical "find the average" problem. It tests for the kind of understanding that the last lessons were working on. Correct answer: **(2)** 81.

Problem 35: GEOMETRY. Your students have seen this problem often enough to breeze through it. If their mistake was in selecting answer choice **(5)** rather than **(4)**, remind them that the parentheses indicate that the operation within must be done first. Ask them to actually carry out the operations in that order, and they will quickly see the error. Correct answer: **(4)** $180° - (90° + 35°)$.

Problem 36: MEASUREMENT. If your students selected answer choice **(4)**, they multiplied both the base and the height by $\frac{1}{2}$. Correct answer: **(3)** 150 cm^2.

Problem 37: GEOMETRY. Students may need a review of the order of the coordinates (x, y). Correct answer: **(1)** *A.*

Problem 38: ALGEBRA. This problem requires manipulation of the distance formula as well as the ability to handle fractional parts of an hour. Proportion would also be used. Correct answer: **(2)** 8 mph.

Problem 39: ALGEBRA. If your students had trouble choosing the correct response using the variable, recommend that they substitute an easy number for x. Correct answer: **(3)** $\frac{500}{x}$.

Problem 40: GEOMETRY. This problem uses all the hints possible to help students. Correct answer: **(4)** $\frac{1}{2}$.

Problem 41: ALGEBRA. The distributive property is the underlying principle of this problem. Correct answer: **(4)** $0.25(55{,}900 - 55{,}433)$.

Problem 42: MEASUREMENT. After eliminating the extraneous information, students can use estimation to solve this two-step problem. Correct answer: **(3)** $20,150.

Problem 43: DATA ANALYSIS. Probability is a topic that is likely to be represented on each form of the test. Correct answer: **(5)** $\frac{1}{600}$.

Problem 44: ALGEBRA. The key to solution is the ability to determine the numbers to use in these complex figures. Correct answer: **(2)** 64.

Problem 45: ALGEBRA. Discuss the various ways that your students divided the area into parts so that they could use the formula. Correct answer: **(4)** 630.

Problem 46: MEASUREMENT. The most critical step is to interpret this as a division problem. The estimation strategy shown in the answer key reinforces the commonsense approach to math. Correct answer: **(4)** 32.

Problem 47: ALGEBRA. In this problem, students are tested both on their ability to use variables and on their knowledge of exponents. Correct answer: **(2)** 21.

Problem 48: GEOMETRY. This classic Pythagorean situation avoids the necessity of actually computing the squares and square roots. Therefore, the problem tests only the placement of the numbers in the formula. Correct answer: **(4)** $w^2 = 16^2 + 12^2$.

Problem 49: ALGEBRA. Estimation is a time-saver on this one. Only one answer choice is less than half of what they now spend, or less than $225. Correct answer: **(5)** $198.

Problem 50: ALGEBRA. Manipulating the interest formula is easier after students have eliminated the time factor. Correct answer: **(1)** $5{,}000 \times .12 \times 3$.

Problem 51: GEOMETRY. To help students who had trouble choosing the correct numbers for the proportion, suggest that they separate the two triangles by drawing them individually. Correct answer: **(1)** $\frac{6}{m} = \frac{9}{30}$.

Problem 52: ALGEBRA. This problem is often difficult for students. When the base is missing in a percent problem, many students lose their intuitive grasp of what is happening. It wouldn't be surprising to find that many found this answer by the "last resort" suggestion—substituting the given answer choices into the problem. Correct answer: **(4)** $300.

Problem 53: NUMBER RELATIONS. Knowing the decimal and fractional equivalents will always be helpful to your students, even on tests. Correct answer: **(2)** 9.6%.

Problem 54: GEOMETRY. As long as students choose the proper formula, this problem is not as difficult as it looks. Correct answer: **(4)** $2 \times 1.5 \times 2$.

Problem 55: ALGEBRA. In this problem, students need to know how to add like variables. Correct answer: **(3)** $6n$.

Problem 56: ALGEBRA. This could be a complicated algebraic equation: $576 = 12(40) + 2(12)x$. Or students could simply unravel it by starting with the parts that are known: $40 \times \$12 = \$480$ is the amount Phil earns in a normal week. Subtracting that from $576 leaves $96 of overtime pay. Finally, divide by $24 to find the number of hours of overtime that he worked. Correct answer: **(1)** 4.

+	0	1	2	3	4	5	6	7	8	9
0	0	1	2	3	4	5	6	7	8	9
1	1	2	3	4	5	6	7	8	9	10
2	2	3	4	5	6	7	8	9	10	11
3	3	4	5	6	7	8	9	10	11	12
4	4	5	6	7	8	9	10	11	12	13
5	5	6	7	8	9	10	11	12	13	14
6	6	7	8	9	10	11	12	13	14	15
7	7	8	9	10	11	12	13	14	15	16
8	8	9	10	11	12	13	14	15	16	17
9	9	10	11	12	13	14	15	16	17	18

X	0	1	2	3	4	5	6	7	8	9
0	0	0	0	0	0	0	0	0	0	0
1	0	1	2	3	4	5	6	7	8	9
2	0	2	4	6	8	10	12	14	16	18
3	0	3	6	9	12	15	18	21	24	27
4	0	4	8	12	16	20	24	28	32	36
5	0	5	10	15	20	25	30	35	40	45
6	0	6	12	18	24	30	36	42	48	54
7	0	7	14	21	28	35	42	49	56	63
8	0	8	16	24	32	40	48	56	64	72
9	0	9	18	27	36	45	54	63	72	81

1	2	3	4	5	6	7	8	9	10
11	12	13	14	15	16	17	18	19	20
21	22	23	24	25	26	27	28	29	30
31	32	33	34	35	36	37	38	39	40
41	42	43	44	45	46	47	48	49	50
51	52	53	54	55	56	57	58	59	60
61	62	63	64	65	66	67	68	69	70
71	72	73	74	75	76	77	78	79	80
81	82	83	84	85	86	87	88	89	90
91	92	93	94	95	96	97	98	99	100

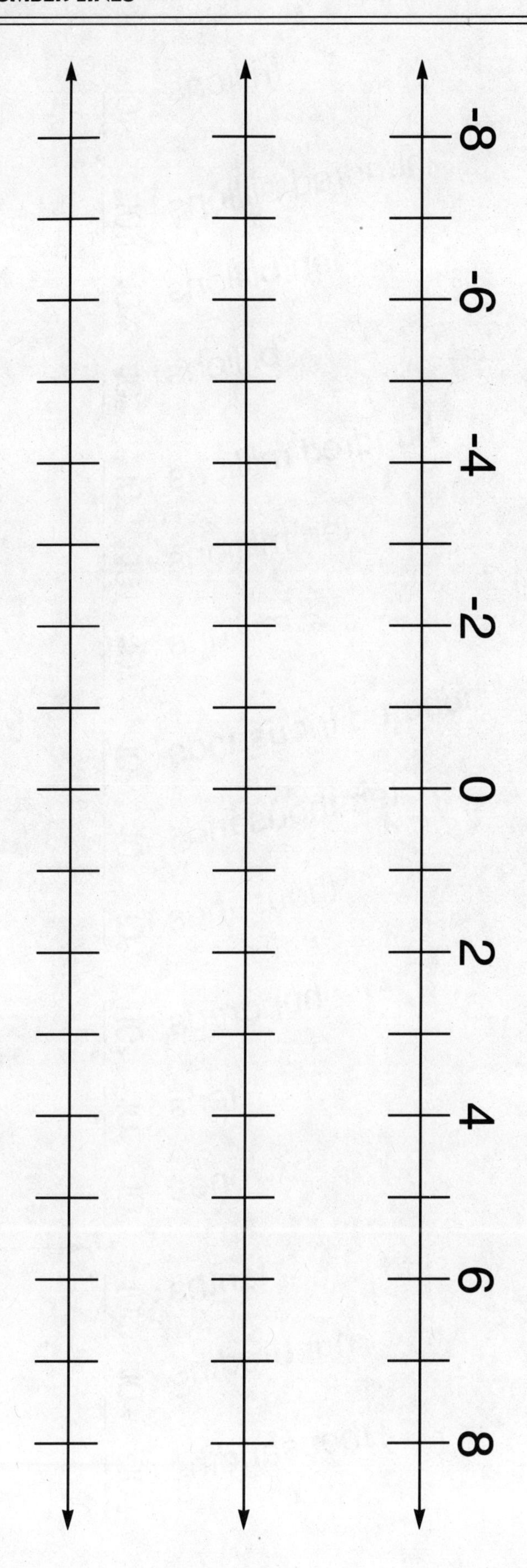
-8
-6
-4
-2
0
2
4
6
8

trillions	10^{12}
,	
hundred billions	10^{11}
ten billions	10^{10}
billions	10^{9}
,	
hundred millions	10^{8}
ten millions	10^{7}
millions	10^{6}
,	
hundred thousands	10^{5}
ten thousands	10^{4}
thousands	10^{3}
,	
hundreds	10^{2}
tens	10^{1}
ones	10^{0}
. ← decimal point	
tenths	10^{-1}
hundredths	10^{-2}
thousandths	10^{-3}

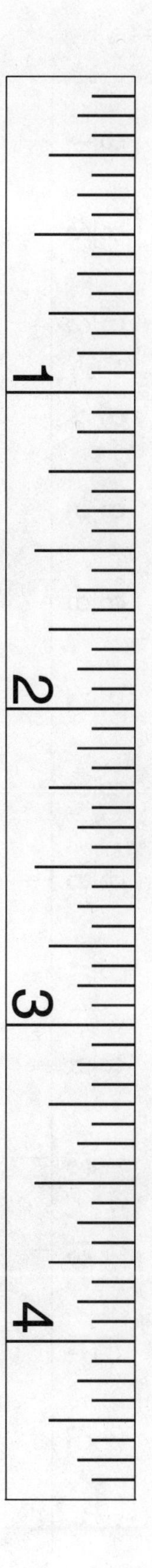
1
2
3
4

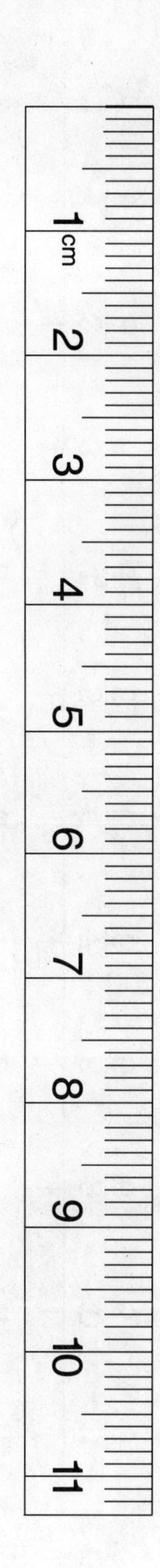
1cm
2
3
4
5
6
7
8
9
10
11

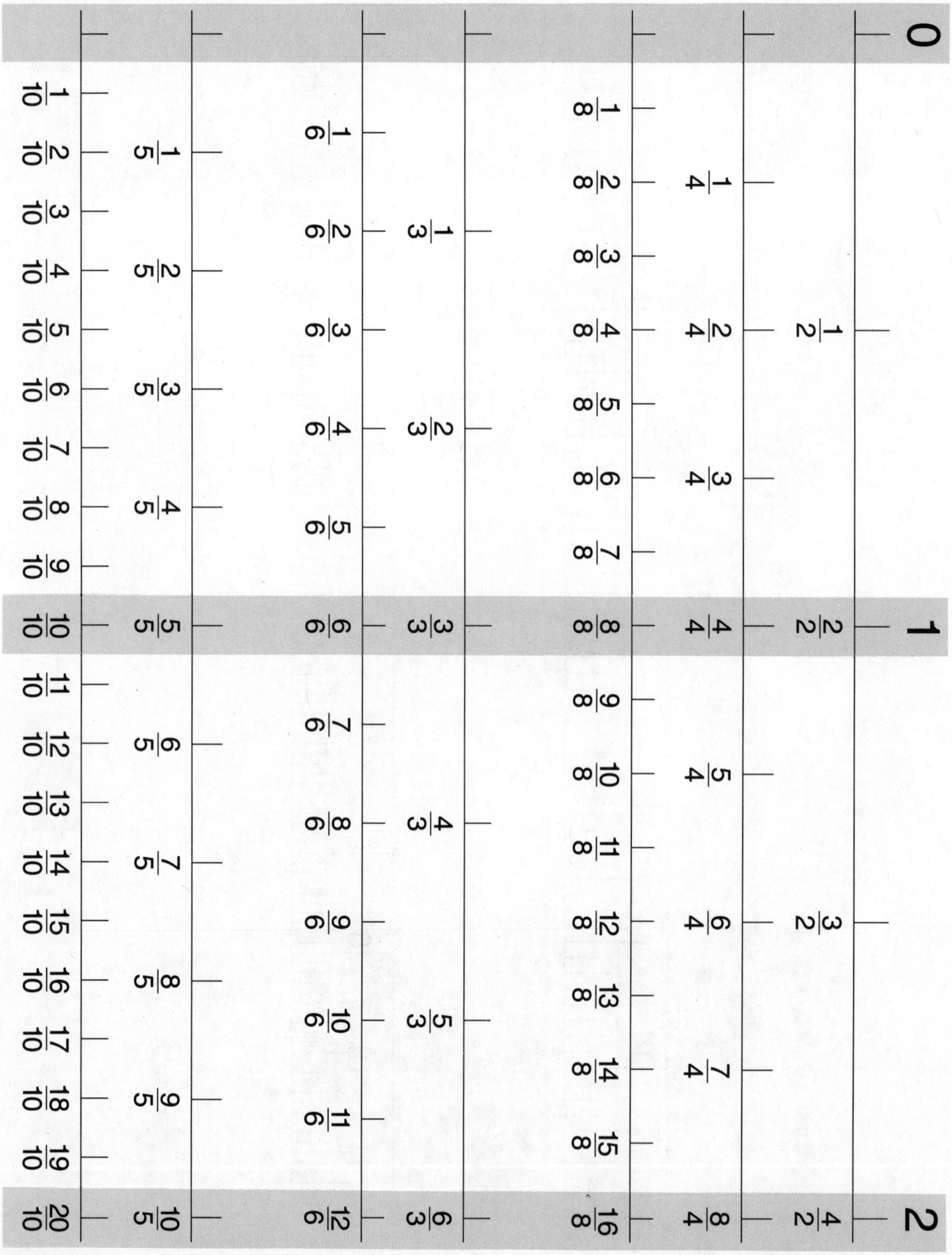

0																			1																				2
									1/2										2/2										3/2										4/2
				1/4					2/4					3/4					4/4					5/4					6/4					7/4					8/4
	1/8			2/8			3/8		4/8		5/8			6/8			7/8		8/8		9/8			10/8			11/8		12/8		13/8			14/8			15/8		16/8
						1/3						2/3							3/3							4/3						5/3							6/3
		1/6				2/6			3/6			4/6				5/6			6/6			7/6				8/6			9/6			10/6				11/6			12/6
			1/5					2/5			3/5				4/5				5/5				6/5					7/5			8/5				9/5				10/5
	1/10		2/10		3/10		4/10		5/10		6/10		7/10		8/10		9/10		10/10		11/10		12/10		13/10		14/10		15/10		16/10		17/10		18/10		19/10	20/10

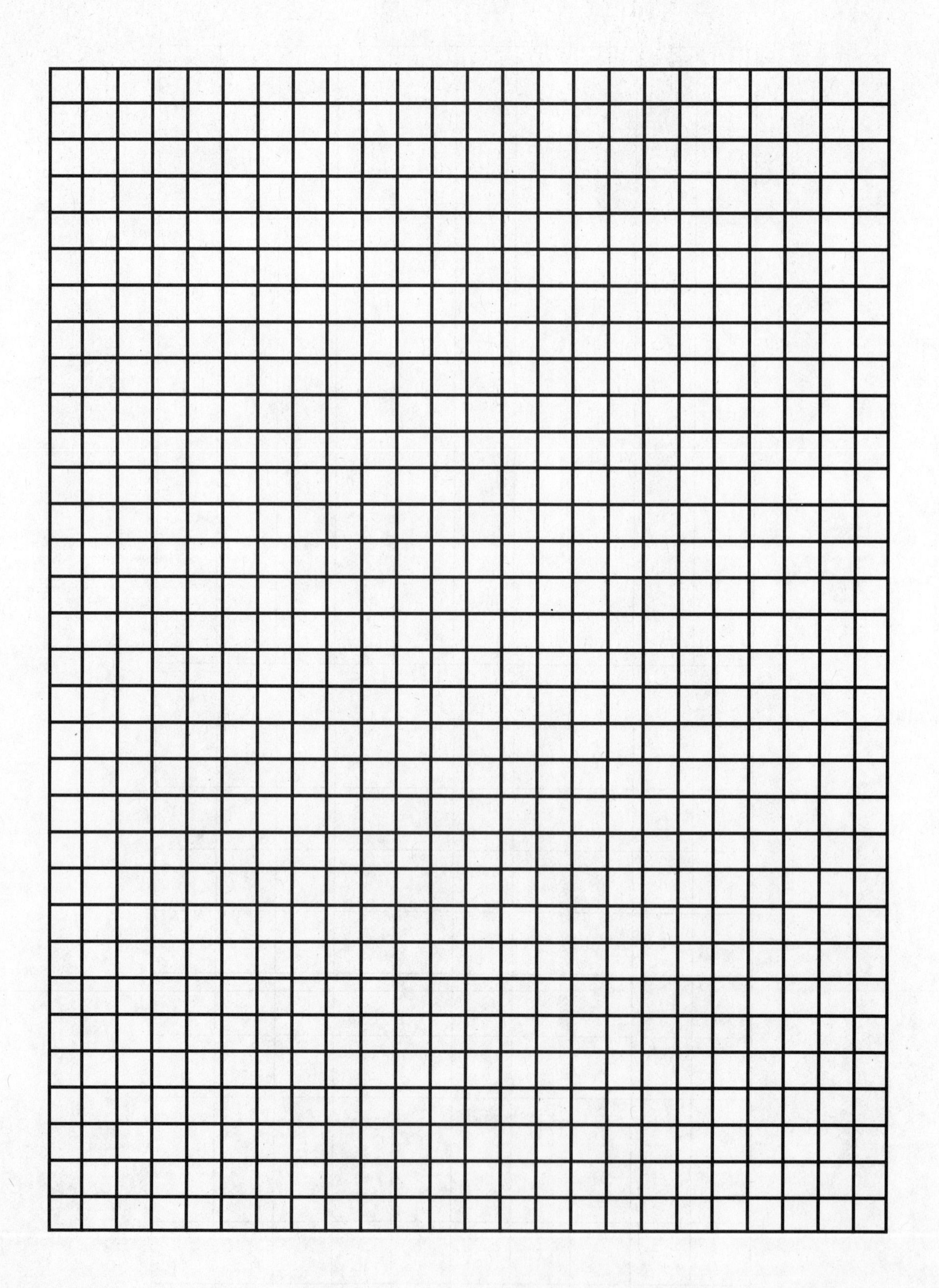